Hydrostructural Pedology

Series Editor
André Mariotti

Hydrostructural Pedology

Erik Braudeau
Amjad T Assi
Rabi H Mohtar

iSTE

WILEY

Preface

This book aims to define as completely as possible the subject matter and main points of this new discipline, hydrostructural pedology, as theorized now for the first time: the underlying concepts, the purpose and role that it has to play within the agro-environmental sciences. It is divided into two parts:

– a theoretical part, where the systemic approach applied to the soil is presented, showing how this leads to the thermodynamic formulation of water in the soil's organized medium and to the systemic modeling of soil–water-coupling in natural or anthropic organizations;

– a methodological part, dedicated to determining the hydrostructural characteristics of a *pedostructure*[1], characteristic parameters of equilibrium state equations and the hydrostructural functioning of soil.

Below we give a brief overview of the key points in the emergence of this discipline up to the development of a physical theory of soil–water; one might indeed wonder and ask why only now, in 2015, is the theory presented as new and complete?

1 If the primary peds form the first aggregation level of fine soil particles, the assembly of these primary peds with other skeleton grains forms the *pedostructure*; it is the motor element of the soil. It forms the main part of a soil horizon (soil layer with a homogeneous structure), with the same hydrostructural properties and share the space with the other sub-systems present in the horizon: roots, biological macroporosity, stones, etc. Its occurrence in the soil horizon and its hydrostructural properties due to mineralogical clays that compose it, first determine the hydral functioning and the agronomic properties of the soil horizon in which it is found.

of the soil; they continue to use the former "water reserve" concepts that are estimated empirically (calculated between two standard fixed retention pressures).

Knowing how to quantify this water and its dynamics in the soil has always been a challenge of great importance in agriculture: it is linked to the water demands of the plant, survival conditions of an ecosystem, resilience in case of climate change, etc. It is only in the new paradigm, where the modeling and the hydrostructural characterization of the soil takes into account the soil structure and its internal organization into aggregates, that this challenge has been overcome. We have recently been able to identify the "green water" of soil to the water of the *pedostructure* (or *pedostructural water*) and thus to model the thermodynamic and hydrostructural properties of this green water using physically established equations (non-empirical). Therefore, we can say that the only soil–water model that takes into account the pedostructure and its hydrostructural properties, including pedostructural water, is the Kamel® model elaborated in the new paradigm of hydrostructural modeling of soil mentioned above.

A radical stance: the natural organization can only be known after its transformation into a system (organized) by the systemic approach

Here we explain why an understanding of the activity mechanisms in the natural environment, or of natural objects such as soils and their physical modeling, requires one to take a clear stance from the beginning with regard to the distinction between organization and system. The systemic approach serves to transform the organization into an organized system available to man to understand its internal functioning, external activity, management, use, etc. We follow the footsteps of Bertalanffy and his companions, who founded the general systems theory (1932–1950), in particular by readjusting the work of the contemporary systemician J.L. Le Moigne [LEM 94] to the issue of the physical modeling of the soil organization functioning with water. A new adaptation has put back the theory of systemic modeling in Cartesian logic, abandoning "the four precepts of the new discourse of the method" proposed by Le Moigne, as the basis for his theory, and replacing them by the four precepts of Descartes that he had refuted "radically".

The discovery of the correct thermodynamic formulation of the soil–water retention curve, and consequently that of the shrinkage curve, would not have been possible without adopting the systemic "Cartesian" approach and that of the new concept proposed in this approach, the SREV: Structural Representative Elementary Volume. The SREV is the base concept of the new paradigm of the physics of water in the organized soil medium. It will replace what lies at the base of the current paradigm of soil–water physics, namely, the REV or Representative Elementary Volume, a concept which was posed as fundamental hypothesis in the physics of continuous porous mediums. Each of the two paradigms has its own system of well-defined descriptive variables of the soil medium, but both are exclusive of one another. Thus, there is a change in paradigm when we assume that the SREV replaces REV, because the system of descriptive variables radically changes. With the SREV hypothesis, we use the set of systemic variables, allowing us to write equations describing the physical processes, and not only, as is the case currently with the REV hypothesis, equations of data using normalized, or averaged variables, which are non-systemic.

The systemic physics of water (gravitational and thermodynamic) in the natural environment constitutes the transdisciplinary language of agro-eco-environmental sciences

The systemic approach of the organized soil medium, as we have redefined by proposing the concept of SREV, logically generates the definition of "natural thermodynamic system" that is closed for the solid phase forming the structure of the organization that it represents, and open to the flows of other phases (water, air) which pass through. The new physics of soil–water that emerges is clearly the transdisciplinary language that allows the interdisciplinary coupling of Kamel® with models of the other disciplines modeling the life and activity of their object of study, which lives in the soil or in association with it. The Kamel® model can in fact be coupled with all abiotic and biotic systems known in the environmental sciences by using variables, equations and parameters of the systemic physics of the water of the soil–plant atmosphere continuum.

Conclusion

The proposed systemic modeling paradigm and the hydrostructural characterization of soil allows for the distinct and quantitative description of three essential properties of soil: 1) the swelling-shrinkage of the soil with the wetting-drying cycles; 2) the coupled dynamics of the pedostructural water (green water) and the gravitational water within the soil; 3) the thermodynamic and hydrostructural equilibrium of the micro- and macro-water couple that constitutes the pedostructural water (water inside and outside of the primary aggregates of the pedostructure).

Acknowledgments

We would like to thank David Pouvreau for his help in proofreading our book and for advising us on various aspects of the General System Theory. A large part of this book, particularly Chapter 6 which offers a historical summary of the evolution of physics over two centuries, would not have been possible without the providential publishing of his thesis on the history of Ludwig von Bertalanffy's general systemology in 2013.

Erik BRAUDEAU
Amjad T ASSI
Rabi H MOHTAR
April 2016

PART 1

Theory of Systemic Modeling of the Pedostructure within the Hierarchal Hydrofunctional Organization of the Natural Environment

The systemic approach to the natural environment is theorized here, in line with work by Bertalanffy and Le Moigne, accurately defining the notion of system with regard to organization, especially the organization of the natural environment. The concepts associated with “black box” or “Representative Elementary Volume” (REV) used across the board in hydrophysics of the natural environment are questioned and should be replaced by the concept of “structural representative elementary volume” (SREV). We have reformulated the systemic approach of the natural environment according to this new concept to adjust it to soil science (defining a methodology of cartography, characterization, and multi-scale hydrostructural modeling of soils), to link the different nested levels of description of the soil organization without conceptual discontinuity between the “soil medium” below and the natural environment at the soil surface. We will show how the systemic approach based on the concept of SREV provides the conceptual means to transform the organization of the soil medium into an *organized thermodynamic system that is closed for solid elements of its internal structure and open for other mobile elements in this structure.* This allows access to a thermodynamic formalization of hydrostructural equilibrium within the soil (distribution of water in the soil structure), which varies according to the water content, and to determine exchanges in the soil matrix (in terms of heat, space, water, air, dissolved matter) with the biological sub-systems living within it. We will also show how the schematic representation of the General System (GS) in its three sub-systems (Operating, Information, and Steering) modified according to Le Moigne [LEM 94], is ideal to represent a scientific discipline in the environmental sciences. In this case, its application to soil science reveals a new discipline, hydrostructural pedology, which will be presented in Chapter 5. The Laboratory of Hydrostructural Pedology, equipped with specific equipment recently developed to meet the data measurement and processing demands of the new discipline is described in this part of the book. It is indeed the focal point of the discipline where the physical characterization of the pedostructure of soils is necessary for their interdisciplinary coupling (bio-physical, agricultural etc.) in situ on the field or in laboratory conditions.

1

Introduction to Part 1

Due to the need to implement sustainable agricultural or ecological systems, with an optimized and respectful use of the environment, the Geographical Information System (GIS) is an essential tool for the management and maintenance of these systems. According to Bordin [BOR 02], GIS is an information system of materials, software and processes, designed for the collection, management, manipulation and display of spatial data to solve planning and management problems. In a more general sense, the term GIS describes an information system that integrates, stores, analyzes and displays geographic information (Wikipedia). To play its full role, the GIS cannot simply be a cartographic database of the physical and managed environment. It must also be a three-dimensional information support, needed for the continuous geo-referenced simulation of the hydric functioning of these agricultural or ecological systems whose soil and water are essential components and resources. This allows managers, operators and other stakeholders of the natural environment to act according to the forecasts provided by biophysical models of agricultural production, water consumption, environmental impact, evolution of the system, etc. [DON 10].

However, to support modeling and simulation, GIS should have *all relevant information* with regard to the physical environment, especially the soil. Is this achievable? What is this relevant information? Soil forms part of the natural soil–plant–atmosphere organization and provides the living and growth space and resources for the plant and all associated biological organisms. The first information layer of the GIS should, therefore, be the mapping units of soils found in the considered zone, similar to the old pedological maps before computers. As we shall see later, there remains the conceptual problem of the *typological definition of soils with regard to their hydric and structural* functioning. Neither pedology nor hydropedology, a recently created scientific discipline intended to solve this problem [LIN 06, LIN 12], has provided a solution.

The lack of a quantitative definition of the characteristics of the hydrostructural functioning of soil associated with the morphological description of its internal organization, prevents *pedological cartography* due to two longstanding practical questions:

1) which methodology should be adopted for the physical (hydro structural) characterization of the internal organization of the pedon, the representative soil volume of a soil mapping unit, and, correlatively,

2) what are the cartographic delineation criteria for these units in the landscape, as defined by their representative pedon?

These two key questions, still unresolved, keep the pedological cartography in a qualitative and empirical description of the soils of an area or region. The soil map is then unusable as an information system to support the physical modeling (not empirical) of the pedon.

Authors of the project SIRSIT-BVM [BRA 01] began to address this issue by creating a Spatial Reference Information System of Irrigated Soils in Tunisia, with the aim of serving as supporting information for the agronomic modeling that would take into account the physical soil properties [BEL 08]. A new methodology for the mapping and characterization of soil based on concepts derived from the *systemic approach and the General System theory* presented by Le Moigne [LEM 94], was implemented. New concepts have been laidout, such as that of *pedostructure* [BRA 01], which is the representative volume of the soil matrix (fabric) in a soil horizon and that of "SIRS-Soils" (Spatial Reference Information System of soils), identified with the Information System (IS) of the General System model (GS) adapted from Le Moigne [LEM 94] to pedology [BRA 01]. An accurate methodology was developed for the hydrostructural characterization of the pedostructure of soils [BRA 04, BRA 05, BRA 06b], which provided the means for a typology of the hydrostructural functioning of soil horizons and thus of soils.

Subsequently, this methodology has continued to be developed by initiating two fields of basic research at the heart of Agro-Environmental Sciences:

1) *the systemic modeling* of natural organizations of the agro-environmental setting [BRA 09a, BRA 09b, BRA 14a];

2) *the physics of hydrostructural and thermodynamic equilibria* of the organized soil medium represented by its pedostructure [BRA 14a, BRA 14b, ASS 14].

These two areas form the theoretical basis of a new paradigm of the characterization and modeling of the hydric and structural functioning of soil proposed by Braudeau and Mohtar [BRA 14a]. As we will see here, the application of this new paradigm reveals a new discipline: "Hydrostructural Pedology", beside the hydro-pedology recently implemented after the mere combination of two original disciplines of the agro-environmental sciences: pedology and surface hydrology [LIN 03, LIN 14].

In this summary, we first define the scientific problems in soil science to show that it cannot progress without a physical definition of the concept of a multi-scale *organized system*, representative of a natural organization (soil). We then show how the systemic approach presented and described by Le Moigne [LEM 94] should be taken and modified to be applied to pedology, the descriptive science of the organization of soil. Finally, we show how the systemic approach, as modified and adapted to soil, leads to the concept of a representative *thermodynamic system* of natural organization, *closed to solid elements of its internal structure and open to other mobile elements in this structure*. This basic concept allows the exact thermodynamic formulation of state equations of the "soil medium" (pedostructure) [BRA 14a]: the soil retention curve (soil water retention pressure as a function of its water content) and the shrinkage curve (apparent specific volume of the soil as a function of its water content). These two soil characteristic curves, well-known in soil science, until now have been represented in soil water models by empirical or semi-empirical functions, which prevents the spatial generalization of modeling results using these functions. The exact thermodynamic modeling of these two curves [BRA 14c, BRA 14d], contrariwise, allows this generalization due to their generic equations and characteristic parameters. Moreover, as we will see in this summary, defining the soil medium as a thermodynamic system, closed only on its solid structural phase (open for the rest), completely opens up the possibility of interdisciplinary couplings between living biological models and the hydrostructural modeling of the organized soil medium represented by the pedon [BRA 14a, BRA 14b].

2

Inherent Problems of Soil Science

2.1. History of pedology

The origins of this discipline are very well defined in the Soil Survey Manual of USDA (Soil Survey Division Staff. 1993), an excerpt of which is cited below:

> "Beginning in 1870, the Russian school of soil science under the leadership of V.V. Dokuchaiev and N.M. Sibertsev was developing a new concept of soil. The Russian workers conceived of soils as independent natural bodies, each with unique properties resulting from a unique combination of climate, living matter, parent material, relief, and time (Gedroiz, 1927). They hypothesized that properties of each soil reflected the combined effects of the particular set of genetic factors responsible for the soil's formation. Hans Jenny later emphasized the functional relatedness of soil properties and soil formation. The Russian concepts were revolutionary. Properties of soils no longer were based wholly on inferences from the nature of the rocks or from climate or other environmental factors, considered singly or collectively; rather, by going directly to the soil itself, the integrated expression of all these factors could be seen in the morphology of the soils. This concept required that *all properties* of soils be considered collectively in terms of a completely integrated natural body. In short, it made possible a science of soil".

Studies gathered an inventory of soil from around the world with this naturalist spirit in the 1970s. Methodologies for the morphological description of soil were set up, confronted the questions of observation scale and typology with regard to the hierarchical structure of the soil and its dependence on its environment, whether field-based (pedological profile, cartographic unit) or in the laboratory (thin-section micromorphology, study of aggregates). The big issue was with the development of a pedological classification that would allow the classification of soils found in a region and the establishment of corresponding soil maps. Faithful to the basic principles of the discipline, the first classifications proposed were pedogenetic [DUC 77] as they were related to a soil typology based on the morphological indices that revealed the key processes of their pedogenesis.

This morpho-pedogenetic typology was essentially qualitative. It allowed the definition of soil mapping units whose characterization criteria are morphology (structure, organization) and morphogenesis (placement of materials, pedogenetic processes linked to climate, to the position in the relief). It also allowed major cartographic inventories on a small scale in the 1960–1980s, but it was unsuitable for large-scale surveys aimed at agricultural development or enhancement, and where soil units must be characterized locally, according to their internal (in the soil medium) and external (on the surface) physics, chemistry and biology. At the time, this was beyond the conceptual scope of the soil science, as we will show below by defining the physical and systemic modeling of the pedon, representative of a soil mapping unit. The deficiency of the pedogenetic approach to large-scale studies contributed to its abandonment by the American classification system, Soil Taxonomy [SOI 75], enforced since 1965.

Being much more pragmatic, this classification system was based on criteria easier to quantify than the morphology and hydrostructural functioning of soil in equilibrium and its dynamics. This makes the definitions of the classes quantitative rather than qualitative and their limits are strictly defined by their physical or chemical criteria and easily measurable on diagnostic horizons. However, these measurable criteria are only indicators or empirical tests, with no physical basis, and thus, their biggest flaw: the chosen limits, though "strictly defined" by the measurable criteria, can only be empirically imposed. This is sufficient to meet methodological standardization needs but certainly not sufficient to allow the physical modeling of characteristic soil processes.

> "For soil survey, the application of quantitatively defined classes to bodies of soil produces quantitatively defined mapping units. This permits the soil maps to be interpreted with more precision than was formerly achieved. Furthermore, this soil classification system simplifies and accelerates the process of soil correlation" [SOI 75].

However, if the soil taxonomy provides ways to quantitatively distinguish soil classes, and therefore better map and establish acceptable statistical correlations, its classification principle does not solve the issue with regard to the hydrostructural functioning of the mapped soil unit. In fact, the quantitative criteria chosen to classify soils are only *indicators* of the physico-chemical functioning of the soil and not the *parameters* of the physical equations (state equations) relative to its hydrostructural functioning, which did not appear until much later on with the concept of pedostructure [BRA 99, BRA 04, BRA 05].

Today the problem is still the same: whether we are referring to the former morpho-pedo-genetic maps from the 1960s–1980s or the current maps from American soil series, these maps closely delineate alleged homogeneous soil units, which are represented by one or several representative pedons (or polypedons). Despite their richness in pedological information, it is still not conceivable, with the soil characteristics that are provided, to access the physical (non-empirical) modeling of the hydrostructural behavior of the pedon at its different internal functional levels, and even less with the modeling of the biophysical activity with regard to the surrounding external environment.

2.2. Modeling of water transfers in the soil: supremacy of pedotransfer functions

In 1999, Bouma *et al.* presented the new situation of pedology regarding the mapping techniques and the presentation of information (GIS, database), while faced with the transformation of agricultural management towards a new paradigm which began to gain importance at the time: "precision agriculture". This was entirely based on the promise of the substantial development of new information and communication techniques; in particular, with regard to the modeling of the soil–plant system and agricultural production systems, which are based on suitable databases (climate, crop system, plants, soil, geomorphology, etc.). To cope with this new demand focusing on modeling, the only possible approach for pedology, according to Bouma *et al.* [BOU 99], was developing what we call "pedotransfer functions": empirical functions obtained by statistical regression on a large number of soil samples according to few basic soil data such as texture (clay, silt, sand), exchange capacity, and organic matter. These functions provide the parameters for the hydrological functions used by soil water models. It was to overcome the lack of knowledge about physical and agronomic properties of soil and especially the lack of accurate data for the scale required; it was also a call for a multidisciplinary approach to the delineation of mapping units.

Today, soil water models are becoming more and more numerous and available on the web. Complete systems of pedotransfer functions have also emerged [RAW 06, PAC 04] to supply these models. However, this method of characterizing and modeling soil-water systems using pedotransfer functions does not allow us to advance in the scientific challenges unique to soils that still exist, namely: change in scale and interdisciplinary coupling of biological or agronomic modeling with soil science [BAL 06], and especially with soil water physics [AHU 06, AHU 07]. This has shown that these issues are challenging to resolve using soil physics together with disciplines whose object of study is dependent on soil. This would require experimental research and the development of new concepts and theoretical models. This is what we will describe below.

2.3. Absence of a unitary theory of the description of soil

Two core problems presented below, intrinsic to cartographic and functional formalization of the natural environment, are the two major conceptual obstacles preventing the creation of any interdisciplinary link between the agro-environmental sciences and soil science.

a) The first problem is knowing how to define and characterize hydrofunctional organization levels of the natural environment.

The soil issue: the question arises, with a different point of view, with regard to its definition, its characteristics, how to represent it, at each organization level (or observation scale) of the agro-environmental environment illustrated in Figure 2.1. This creates as many scientific disciplines, with their objectives, tools, own modeling concepts, as the observation scales being used.

At each organization level identified, the soil appears externally with a form unique to each level and must therefore be described with the qualifiers of this level; the external description, however, must be in line with the internal structure and functionality of the lower level, and so on down to microscopic scales within the "soil medium". This is the problem of scaling.

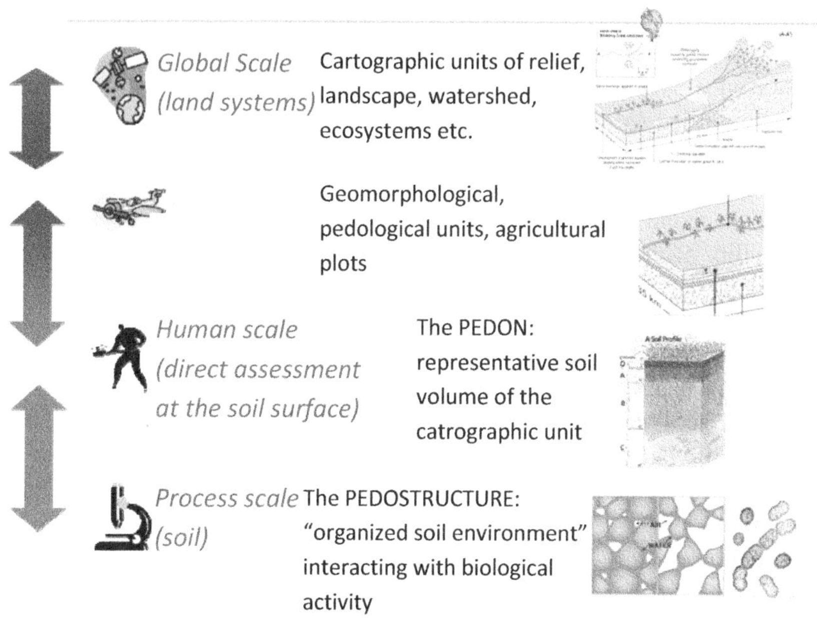

Figure 2.1. *The different observation scales and the different hydrofunctional organization levels of the soil*

The systemic approach can potentially solve this problem by defining the nested, hierarchized organization systems, not only to the soil exterior, from the surface (primary unit of soil included in the geomorphological unit, itself included in the relief unit then the watershed etc.), but also within the soil unit represented by the pedon, which also includes several organization levels (horizons, aggregates, primary particles). Three practical issues arise, for which a response will be given:

– How can we recognize the nested hydrofunctional organization levels then define and delineate the units which belong to each level?

– What are the physical descriptive organization variables corresponding to each functional organization level?

– Also how can we define the physical characteristic properties of an organization unit at a given functional level (for example the pedon)?

These properties expressed at a given functional level should result from the properties of nested sub-organizations defined at lower functional levels.

Later on we will show, after having defined the principles of the systemic approach applied to natural organizations, which response responds to these three questions.

b) The second issue concerns the physical formulation of the hydric functionality of the organized and structured soil medium which is still quite empirical.

Figure 2.2 shows the different natural physical media that clearly stand out in what we call a continuum for water in its bound form: the "soil–plant–atmosphere" continuum. This water is known as "bound", contrary to free or gravitational water which circulates downwards in fissures, cracks and aquifers; bound water is subject to retention forces in the medium, whether it passes through or is consumed (soil-plants, organisms). These mediums have their own non-rigid internal organization and are physically interlinked according to the laws of thermodynamic equilibrium that are still poorly formalized, especially those within soil. They welcome and condition the life of all biological organisms that live there, by exchanging and sharing water between them, as well as space and different forms of matter and energy.

However, the concept of a physical *organized* soil medium, whose hydrostructural thermodynamics conditions life and the evolution of physical, biological, and geochemical processes, is still underdeveloped in soil science. This differs from the thermodynamics of soil water, where the soil medium is seen as an unorganized tri-phasic mixture (solid-water-air), and which was developed in the second half of the previous century (Balbock, Fissel, Low, Miller, Sposito, Parks) up to the 1980s. It was almost completely abandoned after the 1990s for lack of a unified theory of soil water physics that would have been able to consider the hierarchized structure of the soil and its thermodynamic interaction with water. It is based on this water–soil structure interaction that the different physical properties of the soil can be interpreted as shown by Braudeau [BRA 88a, BRA 88b] and Braudeau *et al.* [BRA 04, BRA 05] who called them "hydrostructural properties of soil".

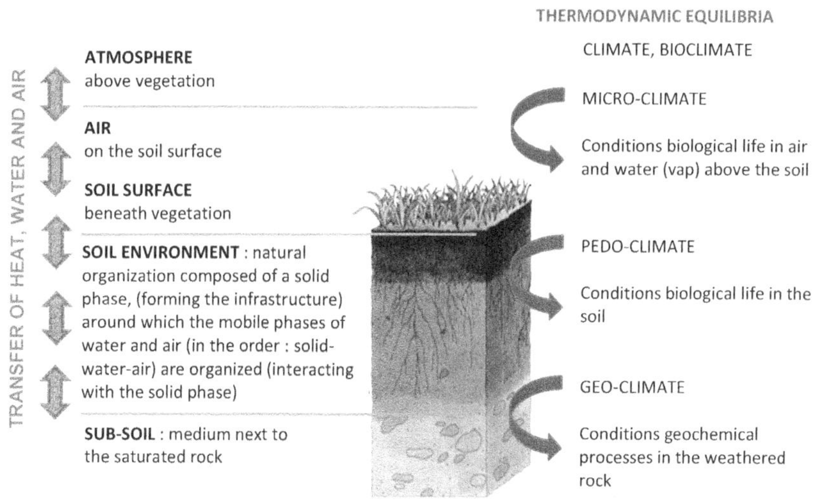

Figure 2.2. *Natural physical media in thermodynamic equilibrium with one another and with living organisms, constituents of the Soil-Plant Atmosphere continuum*

Due to this lack of theory on the water–soil structure interaction, the "soil medium" is the least understood and least studied in the laboratory of the three components of the "soil–plant–atmosphere" continuum schown in Figure 2.2. The soil interior in current biophysical models is modeled as a *black box*, that is to say, a volume occupying space and containing a mixture of three typical phases: solid (minerals, organisms), aqueous and air, without a known recognizable *internal organization*. It is this last point, the fact that no organization between the three phases is seen, which does not allow the internal organization of soil medium to be defined as a thermodynamic system [BRA 09a]; nor to establish exchange relationships, which are thermodynamic equilibrium relationships between the organized soil medium and related systems including the organisms living within.

The two problems that we have just mentioned are the unavoidable challenge preventing the physical modeling (and therefore unitary and transdisciplinary) of natural organizations and of organisms living there.

According to Braudeau and Mohtar [BRA 09a], they cannot be surpassed if the modeling of soil water is kept in the *non-systemic* paradigm of the physics of continuous porous environments applied to the soil medium due to the concept of Representative Elementary Volume (REV) [BEA 72]. In fact, the concept of organized systems, with nested organization levels, organization variables, pedostructure, etc. cannot be defined in such a paradigm where the structure and the internal organization of the soil are masked.

Let us see below how the systemic approach, revisited due to its application to soil science through the concepts of "General System" by Bertalanffy [BER 68] and "modeling" by Le Moigne [LEM 94], allows us to find a solution for these two fundamental problems of pedology.

3

The Systemic Approach Applied to Pedology

3.1. The Bertalanffy project and Le Moigne's general system model

3.1.1. *The general system theory and Cartesian precepts*

Bertalanffy's is the first reference to consider or cite in anything that deals with systemics, the systemic approach and the general systems theory. J.L. Le Moigne directly references Bertalanffy when he named his 1971 pioneering book, now edited several times: "The general system theory, theory of modeling". In introducing the systemic paradigm, he stated [LE 94, p. 55]:

> "This jump that the biologist J. Monod did not dare to do, just another biologist, forty years earlier had already done it; the intuitions of L. von Bertalanffy alone facing, c. 1930, the false debate of the theoretical biology at the time, between an intolerant mechanism and an often childish energy, founded the systemic paradigm: the banner of this gathering bore a name forged by Bertalanffy, the general system theory, whose outline is the same as in this book."

and later [LE 94, p. 60], trying to give an idea of what “the general system theory” is:

> “The general system theory is the theory of modeling objects (natural or artificial, complicated or complex) using this artificial object gradually molded by human thought, which L. von Bertalanffy proposed to call the general system: the system is a model of a general nature (L. von Bertalanffy, in [KLI 72], p. 31). C.W. Churchman defined in one line, in 1964, the nature of this theory: the general system theory is the research methodology of the general system (in Mesarovic [MES 64, p. 175]).”

We will revisit Le Moigne’s book later on as his work on the phenomenological description of the general system theory, according to an original methodology allowing him to comprehensively cover various aspects, allowed us to elaborate our theory of the systemic modeling of the natural environment organization. First, however, we would like to provide some information about L. von Bertalanffy, extracted from the thesis by Pouvreau [POU 13], in the introduction p. 3, which best summarizes the fundamental motivations of von Bertalanffy when giving the title “General system theory” to the research he proposed in 1935.

> “In late 2002, I find myself naturally inclined to read, firstly in French, with often questionable translations, the work of Bertalanffy with the ambitious and promising title mentioned in these essays: General system theory – Foundations, Development, Applications, which I soon began to realize almost invariably refer exclusively to this author. I came across the famous “theory” presented as a response to four phenomena: 1) an increase in the volume of scientific knowledge implying the splitting of disciplines, the specialization of competences and difficulties in communication – especially between “natural sciences” and “human sciences”; 2) a widespread sense of urgency for a solid theoretical framework in non-physical sciences (biology, psychology, economy, sociology, etc.) in order to move beyond the collection of empirical materials controversial with regard to their interpretation; 3) the existence

> in many scientific domains taken from similar epistemological positions consisting in promoting "holistic" approaches to problems posed by "organized complexity"[5] while highlighting their inadequacy, in addressing these problems, of "analytical" thought processes (or "atomistic") often called "mechanistic", deemed characteristic of former physical sciences at the end of the XIX century; 4) the recurring return of certain conceptual models, or even specific mathematical models, in a variety of disciplines with regard to the nature of their subjects. The "theory" developed by Bertalanffy post-war, which he unveiled in 1937 in a seminar at the University of Chicago, claimed to be a framework responding to these findings and adapted to these needs. It was based on the postulate of the possibility of formulating and elaborating principles, models and systemic laws, general in the sense that they would be applicable to various classes of "systems" to be independently defined from the nature of the components of these "systems". Assigned to the "General system theory" were the tasks to actualize this possibility and extract the holistic thought patterns from their traditional confinement to metaphysics to make them accessible to logico-mathematical rigor. A key objective was to exploit "isomorphisms" (similarities between conceptual structures) existing between disciplines and find new ones, and therefore eliminate "superficial similarities" while revealing fundamental deep homologies. It was intended to serve as an outline to build theoretical models in non-physical sciences, and put them onto the right path of the exactitude".

The title proposed by Bertalanffy for his project, known as both general system theory and the general theory of systems, was understood by Pouvreau [POU 13] as "the project of a science of systemic interpretation of reality" to which he gave the name "general systemology", thereby excluding the term "theory", "deemed inadequate and too restrictive". We fully support this appellation that seems to correspond much better to the original intentions. That is to say, a unitarist, but open, framework that was required to answer the fundamental questions arising in scientific research at the time and still today.

Le Moigne is cited among a small number of researchers considered by Pouvreau to have "formed the core of general systemology" and "worked the closest to the development in its various dimensions: Rapoport, Boulding, Rosen and Klir, as well as Mihaijlo D. Mesarovic and Jean Louis Le Moigne".

Certainly, in his book "General system theory, theory of modeling" (1977–1994), Jean Louis Le Moigne described "the general system theory" of Bertalanffy, as it appeared in the 1970s at the time of the first edition, as a modeling theory. He considered systems science as the science of systemic modeling. We can ask whether the theory of modeling (systemic) was still within the framework of the project defined by Bertalanffy and collaborators. This question was actually quite easy to answer: before defining the systemic paradigm, on which the theory of modeling was based, Le Moigne first unveiled the "*liberal and with no philosophical or religious constraints*" point of view which was his modeler. He went on to formulate a new chart for the modeler, composed of four new precepts (Table 3.1), and intended to replace, one by one, the four Cartesian precepts of the Discourse on the Method. According to him, this chart should lead to the constitution of the new systemic paradigm [LEM 94, p. 42]:

> "These four new precepts that we tried to formulate in a condensed form, in the box below, tell us most of the content of the new discourse on the method, to which refers – or can be referred in – contemporary intelligence. We still have to clarify the new paradigm which will form the prototype for this new discourse: we shall recognize the systemic paradigm; then deploy, on this paradigm, a theory of modeling, the General System Theory, which will assist the daily exercise of our modeler intelligence (described as systemic methodology, system analysis or systemography)."

Below (Table 3.1), we list the four precepts of the new discourse on the method outlined by Le Moigne [LEM 94, p. 43] as they mark, in our opinion, the start of the deviation from the stucturalist approach, for which there is no teleological principle explaining the behavior of a natural object, and not "a misleading objectivity" to exclude, since it is this objectivity that is researched in Bertalanffy's "interpretation of reality" project.

The four precepts of the new discourse on the method

The precept of relevance: agreeing that any object we consider is **defined according to implicit or explicit intentions of the modeler**. Never stop doubting this definition, our intentions change, the perception that we had for this object also changes.

The precept of globalism: always consider the object to be known by our intelligence as an immersed and active part within a greater whole. First of all, understanding it, in its functional relationship with its environment **without worrying about establishing a true picture of its internal structure**, whose **existence and unicity will never be taken for granted**.

The teleological precept: interpreting the object not physically, but by its behavior, **without looking to explain this behavior using any law involved in an eventual structure**. However, understanding this behavior and the resources that it mobilizes with respect to projects **that, freely, the modeler attributes to the object.** To identify these hypothetical projects for a rational act of intelligence **and agree that proving them will rarely be possible**.

The precept of aggregativity: agreeing that **any representation is biased**, not due to oversight from the modeler, but deliberately. Consequently finding recipes that guide the selection of relevant aggregates and **exclude the misleading objectivity of an exhaustive inventory of elements** to consider.

Table 3.1. *The four precepts of the new discourse on the method as outlined by Le Moigne in [LEM 94, p. 43] (bold passages highlighted by the authors)*

The above-mentioned four precepts clearly show how the new modeler charter is the exact opposite of that formulated by Descartes. Whereas the latter, where possible, aimed to separate the analysis and description of the natural object from any interpretation or preconceived idea of the observer, Le Moigne, in contrary reinstated this imbrication between the observer and the object observed first by saying (1st precept) that any object (and therefore the natural object, too) can only be defined with regard to the intentions of the modeler. The other three precepts are derived from this: with the modeler's intellectual freedom, we are invited to agree that the desired separation by Descartes is misleading, that we need not worry about the internal structure, nor research an explanatory mechanism of a structure–behavior relationship (precept 3).

It is clear that the four precepts of the new discourse on the method correspond in every respect to the black box paradigm or REV (Representative Elementary Volume) as mentioned earlier (see section 2.3b).

They were proposed by Le Moigne [LEM 94] to justify a *global* (or an external) modeling approach as it does not involve any local (or internal) deterministic mechanisms. This approach is opposite to that proposed by Braudeau and Mohtar [BRA 09a], which is based on the new Structural Representative Elementary Volume (SREV) concept. The SREV, unlike REV and Le Moigne's four precepts, not only recognizes the organization and an internal structure of the object, but also researches the internal physical mechanisms of this organization (hydrostructural interactions), which determine the external behavior of the object with regard to its environment. The concept SREV, therefore, corresponds to the Cartesian precepts and is completely involved in Bertalanffy's project of general systemology as Pouvreau [POU 13] described in his thesis.

These two paradigms are the two poles of the relationship between the local and global description of a zone or region. This local–global relationship is a key theme in environmental sciences: it has never been correctly formalized due to the poor conceptual development of the science concerning the mechanistic functioning of what which forms the local pole, i.e. the soil. This "local-global" relationship is what we propose to "work and deepen" after having detailed hereafter the new paradigm based on the SREV concept and seen its implications in the characterization and modeling of soil water.

3.1.2. *Systemic representation: Le Moigne's two great ideas*

We are indebted to J.L. Le Moigne for the two great ideas that appeared in the original description of the general system theory [LEM 94]. We have taken and adapted them by letting go of his anti-Cartesian vision (and therefore the black box principle) to apply them to pedology: the science of soil organization. These two ideas are developed below:

1) Identifying the description of an organized object in the systemic paradigm by triangulation of the object placed coherently and equidistant between three poles of definition: ontological, functional and genetic (Figure 3.1). Each pole represents one of the responses to one of three questions to identify the object to be described: what it is, what it does and what it becomes? We will transform this representation into a 3D description space defined by three graduated axes each representing the description of the object in response to one of these three questions.

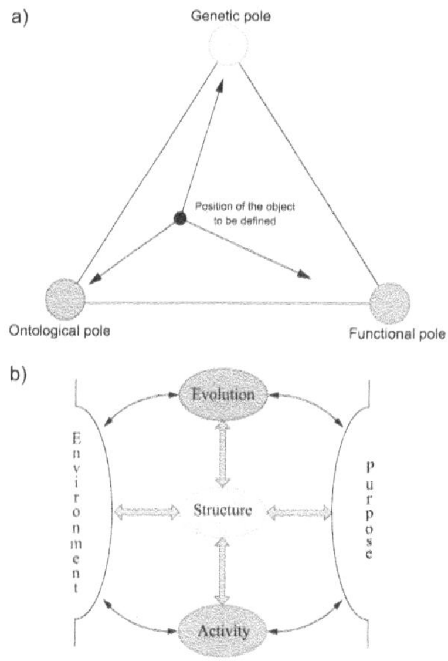

Figure 3.1. *Basic principles of the systemic modeling according to Le Moigne [LEM 94): a) standard definition of an object by triangulation relative to three poles, functional, ontological and systemic; and b) representation of the systemic paradigm*

The systemic paradigm, to which Le Moigne's description [LEM 94] of triangulation methodology refers, is represented by the ideogram shown in Figure 3.1: "An object that, in an environment, with purposes, has an activity and its internal structure evolves over time, without losing its unique identity" [LEM 94, p. 61].

2) A schematic representation of the general system model (Figure 3.2) [LEM 94, p. 148] presented as: "the quasi embryological development of the general system model" borrowed from K. Boulding (by adapting it to our first epistemological manifestations of an instrumental methodology of modeling) appears to have a real pedagogical commodity, as well as a great theoretical generality: we can in fact integrate therein, without distorting it, all classifications of the conceptual systems that have been proposed until now. The "6 level" system ... has a very general symbolic form that allows this articulation *a priori*, giving to the model builder an epistemologically argued starting framework: any systemic model is organized by the matching of an operating system (OS) with a steering system (SS), through an information (or memory) system (IS).

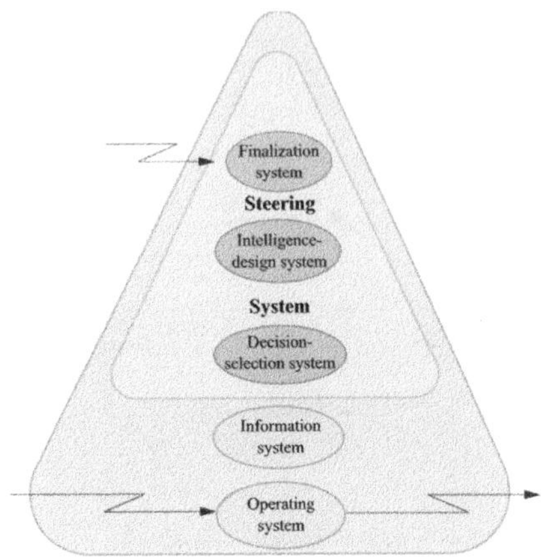

Figure 3.2. *Schematic representation of the general system model in its final evolution/finalization stage (level 9)*

We immediately note that this representation of the general system is organized into three sub-systems and show that these three components, although each is classed as a system, are very different in nature. For example, the operating system (OS) of a natural concrete object cannot be of the same nature as the abstract information system (IS), or the steering (SS) or decision system (DS) that could be replaced by a man or a human organization. However, if we apply to these three different entities the systemic description methodology of an object on three axes of the associated systemic paradigm (Figure 3.1), we obtain a complete and comprehensive description earning the qualifer "systemic", as it defines and describes these entities as systems that correspond to one another. *Now, is the triangulation proposed by Le Moigne in Figure 3.1 valuable for the definition and the systemic description of an object deemed as a "system"?* Here is where the problem lies with the systemic approach: let us see why.

By using each of the three axes of the systemic model, a systemic description of any *man-made* system whether SS or IS is always possible. However, this is not the case for a natural system, in particular an OS of the GS, which is a system sampled within a natural organization *for which the functional boundaries (of sub-systems) of the internal organization are unknown.* The organization of which OS is representative is usually defined on two axes, "ontological" and "genetic", but the functional axis is not clear.

This is the main problem with soil mapping in pedology: how can we define the internal hydrostructural functioning of soil in a manner that differentiates it from its neighbor and how can we delineate its spatial extension? This was to address the issue with the mechanistic description of how the soil structure functions with water on the functional axis as the REV concept had been established early in soil hydro-physics [BEA 72]. It is still used today in models of hydric functioning of soil. This REV concept is similar to the "black box" concept that is particularly advocated by Le Moigne, but it is opposite to the concept of the system and, *a fortiori*, of the "organized system", thus preventing the development of this key concept as we will show later on.

The problem, therefore, lies today in the systemic description of OS (operating system of the general system): the object to be modeled, including its internal organization and functionality (internal and external), must be describable in the standard triangulation shown in Figure 3.1(a). It must be consistent and in line with the associated systemic paradigm (Figure 3.1(b)) in order to be accepted as an OS of the general system (GS, Figure 3.2).

Thus, *provided this condition is added to describe the OS on the three systemic description axes*, then, the general system model described by Le Moigne [LEM 94] (Figure 3.2) is just as Bertalanffy predicted: a generic system to develop knowledge of the natural world and in which any scientific discipline can be identified or referred.

How, then, can we systemically describe the organization of soil according to the 3 poles of reference shown in Figure 3.1(a)?

3.2. The systemic description of the soil organization

3.2.1. *Physical definition of a "system"*

It is important to define the concept of a system compared with that of a natural organization. In principle, a system is entirely formulated and controlled by a person who has a detailed understanding of the external form, the material composition of internal organs (the sub-systems) and the internal functioning at all organization scales (organs and assembly): the cause of the overall activity of the system relationship with its external environment. This understanding of the system with regard to its internal functioning allows the control of its activity with respect to the exterior, as represented in the GS in Figure 3.2. On the other hand, when it comes to a natural organization, the person only has, more often, an empirical knowledge of what appears externally to the object of study when considered at a given organization level (e.g. the soil surface, at the level of the plot).

The person then has the choice between two scientific options: 1) to stop at the empirical investigation stage of the natural environment (naturalist) and put into practice the knowledge acquired at this stage (i.e. empirical) to exploit the natural environment; or 2) to continue the investigation on the interior of the natural object to understand its physical relationship with the external activity. The second option leads to the development of a theory to explain the physical link between the inside and outside of a natural organization, and therefore the validation would agree with empirical knowledge of the object used in the first option.

However, in both cases, the first step is the artificial demarcation of the object to be used or studied, *which transforms the "piece" of the natural organization considered in a closed system for its internal structure.* Physically speaking, this demarcation only addresses the structure of the medium, organized and positioned in space, in which air, water and living organisms move. It therefore contains a determined fixed quantity of solid matter (the structure) but is still completely permeable to flows of mobile elements (in particular air and water). This operation is not neutral; it discriminates between the solid phase forming the structure of the object and the liquid and gas phases contained in the object.

The outer boundaries that we assign to an item in the study, thus determining the volume occupied by this object (e.g. a soil sample), is therefore part of the system definition.

It is curious to note that the definition of a system in the current literature[1] does not include this absolute necessity for the external boundary

1 This does not mean that there have not been any studies in systemics on the system boundary issue, a crucial problem faced by Bertalanffy (Hall & Fagen, 1956) which occupies many studies related to "critical systems theory". I simply wanted to point out that in current scientific language, which uses the term system on any occasion, we cannot find the definition of a system that concretely mentions the external limit. This boundary, impermeable to the solid phase of the structure but transparent to all other flows of matter, will allow the categorical differentiation between the organization and the system and deem the REV not a system unlike SREV.

of the system, which allows us to theoretically determine the fundamental variable of any system: its specific structural volume $\bar{V}_{tot} = V_{tot}/M_s$, whose structural mass M_s enclosed in this volume is taken as a reference, as it is fixed and positioned in space.

Here is how the concept of a system is presented on Wikipedia ("Systems thinking", http://en.wikipedia.org/wiki/Systems_thinking):

"Science systems thinkers consider that:

– a system is a dynamic and complex whole, interacting as a structured functional unit;

– energy, material and information flow among the different elements that compose the system;

– a system is a community situated within an environment;

– energy, material and information flow from and to the surrounding environment via semi-permeable membranes or boundaries;

– systems are often composed of entities seeking equilibrium but can exhibit oscillating, chaotic or exponential behavior.

A holistic system is any set (group) of interdependent or temporally interacting parts. *Parts* are generally systems themselves and are composed of other parts, just as systems are generally parts or holons of other systems".

The concept of the *system boundary* is not mentioned. It must be understood that the physical variable $\bar{V}_{tot}$ born from this boundary setting (or discretization of the natural medium) and which we have seen is fundamental for the system concept, however does not exist in the set of the descriptive soil variables required by current soil–water models, as we shall see later on.

We then propose this physical definition of the *natural system*: any natural organization that man has defined in its spatial extension by boundary setting (splitting, boundary) externally based on the structure (or infrastructure) of this organization that it encapsulates. Once defined, the natural system can then be considered as an operating system (OS) of the general system (GS) representing the following scientific discipline: i.e. in charge of study of the OS and whose scientific product is the information system (IS) describing the OS according to the systemic reference frame of three poles (Figure 3.1). However, to establish this information system, like the internal organization of the OS, it is necessary to study its internal organization (option 2 of scientific investigation). For the considered organization to not only be a system (which is due to its external boundary) but also to be an *organized system*, its internal organization should be exhaustively partitioned into sub-systems (abiotic or biotic), closed only on their solid component (remaining open to flows of water and air), so that their volumes are accurately determined, and the sum exactly corresponds to the overall volume of the operating system. A natural organized system must therefore have; 1) a concretely drawn external boundary that accurately determines its interior and exterior; and 2) an internal volume V_{tot} partitioned into sub-systems of volumes V_i whose sum exactly adds up to the total volume: $V_{tot} = \sum V_i$.

Living organisms that are naturally delimited by a solid natural boundary (membrane, skin, carapace, etc.) are full systems and share the poral space of soil with other sub-systems within the total volume of the defined soil system.

The question now is how to describe the internal organization of the OS (of the corresponding GS, e.g. the soil studied in pedology) in the triangulation proposed by Le Moigne in Figure 3.1, by using the three poles of description: ontological, genetic and functional?

3.2.2. *Graduation of spatial axes: the systemic description of soil*

To describe and define a natural object in our physical world, we return to Le Moigne's idea of "triangulation of the object to be described" (Figure 3.1(a)). However, we have replaced the three poles (ontological, genetic and functional) of the standard triangulation with three graduated axes, each representing one of the three fundamental and inseparable aspects that all natural objects have. This object has a form (organization) and substance, it evolves, transforms over time, and is active, acting and reacting in relation to its exterior.

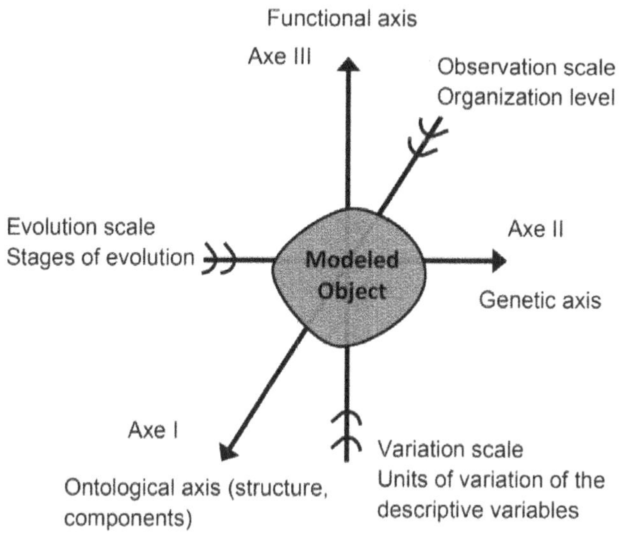

Figure 3.3. *Graduation of the three descriptive axes for definition and modeling of soils, adapted from the three descriptive poles in the triangulation according to Le Moigne [LEM 94]*

These axes, shown in Figure 3.3, may be called ontological, genetic and functional just like the corresponding poles, but other names that describe these three aspects can also be used based on the object under study. With regard to soil, they can be called: organizational axis (matter and form, or structure and organization), pedogenetic (pedogenesis, transformation phases of soil units) and hydrofunctional (physical properties of soil in response to its external environment).

With regard to the axes graduation, Le Moigne considered the succession of stages of evolution on the genetic axis (different levels or stages of evolution of a system), but not the ontological axis (structure-organization) as being graduated into hierarchical organization levels to be adapted to the description of a natural object, such as soil. This has been morphologically described by naturalists like Brewer [BRE 64] as an object *structured into organized and hierarchical elements, with several organization levels.*

This graduation of axes is necessary for the systemic description of the organization of the SO of the GS, as it reveals the hierarchy of organization levels on the 1st axis (ontological) and succession of soil evolution stages on the 2nd axis (genetic). Regarding to the 3rd axis, specifically dedicated to the quantitative description (modeling) of the internal and external physical properties of the soil, it is the axis that physically models the hydro-structural functioning of the soil at different organizational scales recognized and defined on axis I. The multiple graduations of axis III are those of descriptive variables defined on axes I (organizational variables) and II (evolution variables) and put into equations on this axis.

Therefore, a pedological coverage is described on the three axes follows:

– On the organizational axis I, it is the hierarchical spatiality of the internal organization of the pedological coverage that should appear in the descriptions. Classical pedology long ago developed the methodology to describe a soil profile at different organization levels. The aim is to identify the different nested hydrofunctional organization levels which emerge by themselves at the soil surface, i.e. in the landscape; also in the soil represented by the pedon (Figure 3.4).

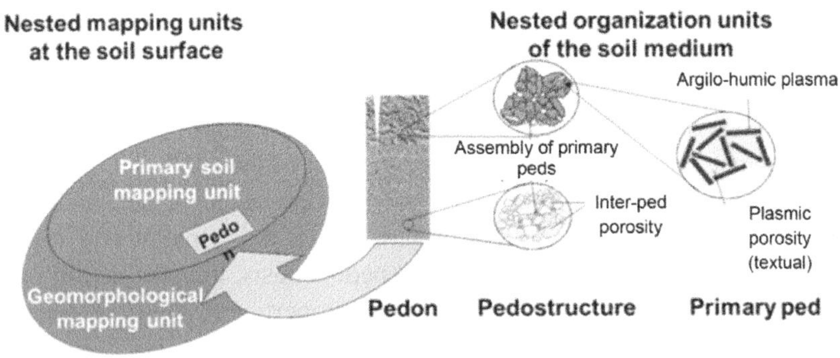

Figure 3.4. *Schematic representation of the internal and the external functional organization levels of soil qualitatively described by pedologists (taken from Braudeau & Mohtar [BRA 09a]). For a color version of this figure, please see www.iste.co.uk/braudeau/hydro.zip*

The "systemography" of the object to be described (term used by Le Moigne [LEM 94]) involves delineating or defining the volumes of the different organizations belonging to the same hydrofunctional organization level (see Figure 2.1): the soil mapping units, for example, are thus presented as operating systems.

– On the genetic axis II, the graduations reveal the stages of evolution that follow each other. At a given moment, and therefore at a given stage of evolution, the system is described relative to the two other axes (structure and functioning). Describing the evolution of the OS on the genetic axis ultimately comes back to tracking the hydrostructural and -functional characteristics of its organization as a function of time in different stages of evolution on axis II. By describing the soil organizations on this axis, naturalist pedologists reveal the processes of pedogenesis, degradation, transformations and stages of evolution, which are used as criteria in the classification of soils.

– The functional axis III is the axis that models, or describes in physical and mathematical language, the dynamic functioning and internal and external activity of the organized objects in response to activities and events in the environment. The extensive variables used in equations of functioning and activity will have been previously defined on the organizational axis I where the organizations and assembly levels are recognized. The graduations on this axis are those of descriptive variables in their domains of variation.

Note that the descriptive variables characterizing the sub-systems defined on axis I (organization-structure) are inevitably those found in evolution equations defined on axis II (pedogenesis-evolution) and process equations defined on axis III (hydrofunctional). The systemic definition of describable variables on axis I and the use of these variables on the other two axes is what will assure the consistency and completeness of the description of the OS, a description that will be fully reproduced by the information system of the corresponding GS. These inter-axes relations are described in Chapters 7 and 5 of Le Moigne's book [LEM 94] where they are listed and described but, of course, without the "organized system" concept responsible for the overall consistency, being physically defined as above.

So this description space acts as a mathematical operator that transforms the natural organization, with undefined boundaries (unknown), into an *organized system* with physically defined and quantifiable internal and external boundaries. This helps link the qualitative description of natural organizations and sub-organizations (axis I) to the quantitative description of their characteristics (axis II) and their hydrofunctional properties (axis III). This is what we mean by the "*systemic description of the organization of a natural system*".

3.2.3. *Systemic modeling of the operating system (OS) on axis III*

Axis III is, therefore, the site of the descriptive quantitative (modeling) of the internal activity (hydrostructural) of soil produced by the interaction of the soil structure with the flows of matter (water and air) and energy, and by the exchanges with the exterior. This internal activity of soil obeys natural specific physics of the "soil medium" as explained on axis III using the appropriate set of physical variables from the OS systemography (into organized sub-systems) on axis I.

However, as noted by Braudeau and Mohtar [BRA 09a], today there is a fundamental disconnection between hydrofunctional modeling of soil on axis III and the pedogenetic description and its functional organizations on the plane of axes I and II: the two corresponding disciplines, soil hydrophysics and pedology, coexist independently when they have the same object of study – the soil. It is this disconnection between the two disciplines that we show in Figure 3.5.

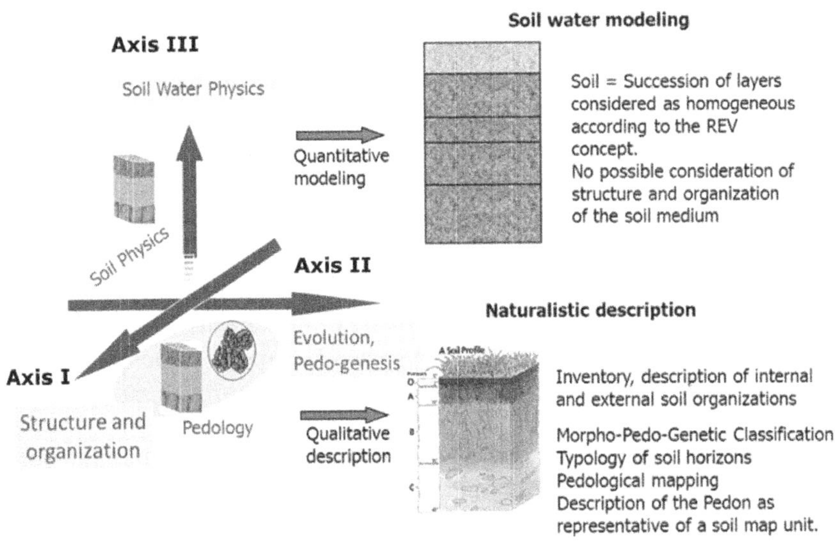

Figure 3.5. *Conceptual breakdown between soil water physics (axis III) and pedology, the science of the morphological and the pedogenetic description of soil organizations (plane of axes I and II)*

In fact, except for the Kamel® model that we will present below, all models of the physical functioning of soil with water are based on a very reductive assumption about the homogeneity of the porous soil medium, the REV hypothesis (Representative Elementary Volume, Bear [BEA 72]), in order to use the equations from the mechanics of continuous porous environments. Thus, the descriptive variables of the soil medium used in these models are defined according to the REV concept. This concept is similar to that of the "black box", seen above as a modeling option in the cybernetic paradigm considered by Le Moigne [LEM 94, p. 54]: as with REV, the internal structure is not determined (substitution of the mechanistic structure–behavior point of view by the external or global point of view: behavior–purpose). The use of the REV concept therefore makes defining descriptive variables of the internal organization of the porous environment studied impossible, and causes this complete disconnection of axis III with the other two axes, as shown in Figure 3.5 [BRA 09a].

In fact, the representative elementary volume (REV) is not a system, as previously defined. It has no accurately defined boundary according to the internal structure that it encloses. It cannot, therefore, be described in the systemic model with the three axes. As the extensive variables of the REV (e.g. water content) cannot be related to the structure, they can only be reported to its volume which, however, is not concretely defined. This creates a set of volumetric variables, such as density (ρs, mass of the solid phase in its apparent volume), porosity (V_{void}/V_{total}) and volumetric water content ($\theta = V_{water}/V_{total}$) which are macroscopic variables. They are averaged over the total volume considered, not localized and not describable on axis I. They can therefore not be used in the process equations of axis III.

On the other hand, the SREV (Structural Representative Elementary Volume) proposed by Braudeau and Mohtar [BRA 09a] to replace REV, is a system: its volume exactly corresponds to that occupied by its structure and the descriptive variables of the internal organization of the SREV are variables of the organization (extensive) linked to the structural mass, M_s, the structure mass of the SREV. These are localized variables with regard to structural elements and are therefore featured in process equations. Replacing the REV with SREV in soil water physics made it possible to establish a set of systemic descriptive variables of the organization of the pedon (axis I) entirely adapted to its precise quantitative description (modeling) on axis III. It was followed by the writing of the computer model Kamel® [BRA 06a, BRA 06b], which models the hydrostructural functioning of the pedon at different organization levels and in perfect correspondence with elements of its internal organization defined on the plane of axes I and II as shown in Figure 3.6, where axis III now rejoins the others. Figure 3.6 also shows an example of the change caused by replacing the concept REV with the SREV when writing the equation of water transfer in soil. The variables used are defined according to the SREV concept: the gravimetric water content W and the apparent specific volume $\bar{V}_{ps}$ of the pedostructure (structure of the soil matrix defined by Braudeau *et al.* [BRA 04] are systemic variables related to the mass of the pedostructure forming the SREV. In the equation, they replace the variables θ and ρ_s which, as we have seen, are non-systemic (variables of REV related to its materially undefined volume).

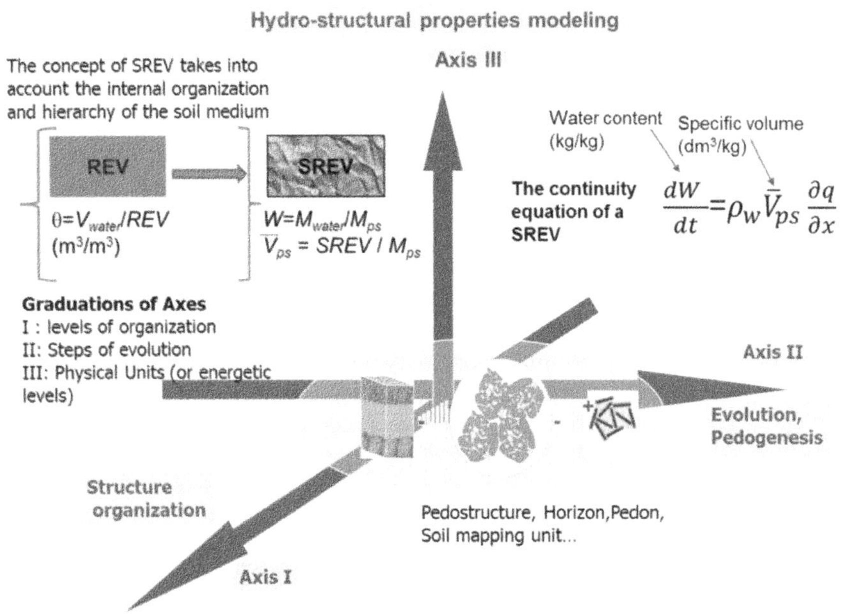

Figure 3.6. *The SREV concept allows the junction of the three axes of the systemic description of the same organized soil system, and in particular the modeling of its hydrophysical activity at different organization levels and its different degrees of evolution. For a color version of this figure, please see www.iste.co.uk/braudeau/hydro.zip*

Moreover, the two variables W and $\bar{V}_{ps}$ are state variables that explicitly feature in the water transfer equation represented in Figure 3.6 (where q is the flow in m/s, t is the time in seconds and x is the depth in m). For comparison, the transfer equation used in the current soil–water models, which are based on the REV concept, is written as: $d\theta/dt = \partial q/\partial x$, where only the normalized state variable θ is used $\left(\theta = W/\rho_w/V_{ps}\right)$.

It is clear that unless we use systemic variables (defined according to SREV) descriptive of the internal organization of the soil volume studied (such as the OS of a GS), it is impossible to model the hydrostructural properties of soil like with the Kamel® model. This takes into account the variations in volume (shrinkage–swell) of the internal organizations (pedostructure, primary aggregates) with the water content, equilibria of water potential in the micro- and macroporosity of the pedostructure, etc., which cannot be considered by models built in a non-systemic paradigm like those based on REV.

3.2.4. *The "Structural Representative Elementary Volume" (SREV) concept required for the systemic description of the pedon*

3.2.4.1. *Definition and characteristics of the SREV of the matrix medium of a soil horizon*

The SREV is the volume of a section of the soil medium that it represents, such as a soil sample whose internal structure will be conserved (for study). The structure of a soil SREV is sectioned by discretization (or sampling), such that the structural mass of a SREV remains constant in relation to the volume. Liquid phases and air are not disturbed by this hypothetical division and will continue to freely circulate through. It is thus a thermodynamic system, closed to its structural solid phase and open to water and air. An undisturbed sample in a soil horizon can be used as a sample of the pedostructure of this horizon: a tri-phasic medium formed as an assembly of primary aggregates and coarse particles (coarse sand, nodules, gravel etc.) in variable amounts.

The characteristics of this representative structural volume (SREV) are as follows:

1) its boundary is concrete in the sense that, once drawn during the discretization of the medium, it always takes account of the same constant mass of solids that make up the structure. This envelope is impermeable to the solid phase, unlike the REV, and remains permeable to the liquid and gaseous phases mobile in this structure;

2) all variables defined on the SREV are no longer related to the variable volume of the SREV but to the mass of the structural solid phase limited in this volume. The variables, unlike those of REV, are physically defined with respect to the internal and hierarchical organization of the "soil medium". A sample of the soil pedostructure is an example of the SREV of the matrix medium of a soil horizon.

This possibility to transform the organization of the soil medium into a system through the SREV concept has a crucial consequence on the soil physics: the concepts of the structure and hierarchical organization of the "soil medium" can now be included in the physical and thermodynamic equations of soil–water models which will be used to describe the hydro-structural functioning of the pedon in the systemic paradigm of the three axes of description presented in Figure 3.6 [BRA 14a].

3.2.4.2. *Internal organizations of the pedon*

With these two concepts of SREV and of the systemic description model, the different hydrofunctional organizations within the soil and at the surface can be determined and defined on the plan of axes I and III (Figure 3.6).

Thus, the soil mapping which is used for the determination and characterization of soil units in the landscape, can be made, as shown in Figure 3.4, in connection with the internal organization and the hydro-structural functioning of the representative pedon.

The pedon is the volume of the soil (Figures 2.2, 3.4 and 3.7), usually with an area of 1 m^2 and a depth of 1.20 m, which is considered as representative of the soil unit to which it belongs. A visual observation of the terrain (pedological pit) defines the pedon horizons, which are mainly differentiated by their structural morphology (type of structure, color, porosity, etc.).

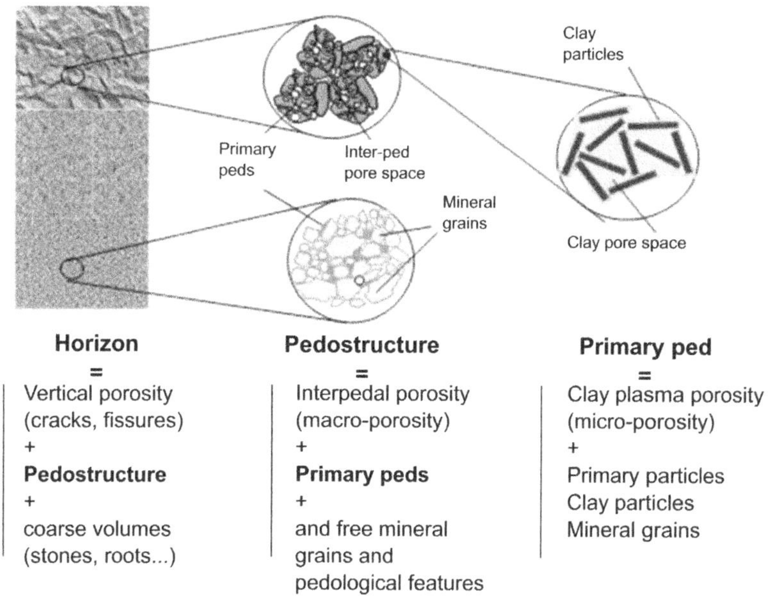

Figure 3.7. *Schematic representation of the pedon and its internal organization*

Using the SREV concept to define the systemic variables, describing in detail the pedon and its internal organization, allows the physical modeling (non-empirical) of the hydro-structural functioning of the pedon at its different nested organization levels represented in Figure 3.7 (pedon, horizon, pedostructure and primary aggregates). This new paradigm of hydro-structural modeling of soil has become a reality with the Kamel® computer model entirely besed on the SREV concept [BRA 09b, BRA 14a].

This modeling of the soil horizon was initially considered to be solely composed of the pedostructure, i.e. the matrix material of the soil only, without considering the other biotic and abiotic components (roots, biological macropores, pebbles or other coarse elements) that may be present in a large amount in a soil horizon. The corresponding set of systemic variables that has been implemented to describe the pedon, its horizons and hydrostructural properties of the characteristic pedostructure making up each horizon, is presented in Table 3.2. The initial versions of the computer model Kamel® were developed based on the systemic description of the pedon, as shown in Figure 3.7, whose descriptive variables are given, for each horizon, in Table 3.2 [BRA 06a, BRA 06b, BRA 09b].

Functional level of the pedon	Specific volume (dm³/kg)	Poral volume (dm³/kg)	Water content (kg/kg)	Water retention pressure (kPa)	Hydric conductivity (dm/s)	Non saturated water (kg/kg)	Swelling water (kg/kg)
Horizon SREL		$\overline{Vp}_{fiss}$	W_{fiss}	h_{fiss}	k_{fiss}		
Pedostructure	$\bar{V}_{ps}$		W_{ps}	h	k_{ps}		
Interpedal pore space		$\overline{Vp}_{ma}$	W_{ma}	h_{ma}	k_{ma}	w_{st}	w_{ip}
Primary aggregates		$\overline{Vp}_{mi}$	W_{mi}	h_{mi}	k_{mi}	w_{re}	w_{bs}
Primary particles	$\bar{V}_s$						

Table 3.2. *Descriptive variables of the hierarchical internal organizations of the pedon*

This table shows a comprehensive list of state variables of the *pedostructure* and its organizational sub-components, as if it completely filled the soil horizon. Using this set of variables (which we can now deem systemic) allows us to study the hydrostructural functioning of the pedostructure in the laboratory, and to establish the systemic equations of the different hydrostructural properties of the pedostructure: shrinkage curve, water retention curve and swelling rate of aggregates in water, and hydric conductivity curve [BRA 04a, BRA 04b, BRA 06b, BRA 09a].

In the following section we present the new soil water physics, which was not possible until the REV concept was replaced by SREV [BRA 09a] leading to the physical definition of a *thermodynamic system* of the pedostructure and, further (3.3.5.3), of soil layers created by the systemic discretization of a soil horizon.

3.3. Systemic physics of the organized soil medium defined on axis III

3.3.1. *The thermodynamic system of the pedostructure*

The pedostructure, as shown in Figure 3.7, is defined as the tri-phasic organization (solid, aqueous solution and air) of the soil medium into the primary aggregates [BRA 04], in line with the naturalist's description of the hierarchical soil structure by Brewer [BRE 64]. However, this organization of the soil medium is not recognized in the current modeling based on the REV principle, whether in thermodynamics or soil hydro-physics, where the tri-phasic soil medium is processed (modeled) as a homogeneous mixture of three phases: no distinction is made between the phases and thus the structure of the solid phase cannot be taken into account nor the specific organization of the three phases between them.

However, in addition to the soil arrangement into the primary aggregates, we must also take into account the other feature of the tri-phasic organization of the soil: the ordered arrangement of the three phases in relation to one another. The solid phase is surrounded by the liquid phase, which is surrounded by the gaseous phase. Solid particles are in direct contact with other solid particles or with water, but never with air in a living soil. The consequence of this ordering in soil water physics, and particular in thermodynamics, is that there are two interfaces to highlight and consider for the definition of descriptive variables of the organization: solid–liquid (s.w.) and liquid–air (w.a.) which are the inner and outer boundaries of the water layer surrounding the solid phase. These two boundaries exactly define the *thermodynamic system of water in the pedostructure* and associated state variables (systemic): $\bar{V}_w$ and W (Table 3.3).

In addition to this arrangement of phases, whose corresponding organizational variables are listed in Table 3.3, we should also take into account the structure of the solid phase in the primary aggregates of the pedostructure. Primary aggregates (or primary peds) are made up of clay plasma formed of clay particles and fine silt particles, which may also contain few other coarse particles (quartz or other). They are termed *primary* because they have no trace of internal fissures, indicating that they are not composed of several sub-aggregates [BRE 64].

Phases	Mass	Pedostructural volumes	Water and air content	Specific volumes
Solid	M_s	$\bar{V}_s = V_s/M_s$		$\hat{V}_s = V_s/M_s$
Liquid	M_w	$\bar{V}_w = V_w/M_s$	$W = M_w/M_s$	$\hat{V}_w = \bar{V}_w/W$
Gas	M_{air}	$\bar{V}_{air} = V_{air}/M_s$	$\bar{A} = M_{air}/M_s$	$\hat{V}_{air} = \bar{V}_{air}/\bar{A}$

Table 3.3. *Descriptive variables based on the distinction of three ordered phases of the pedostructure*

Two hydrofunctional sub-systems of the pedostructure are distinguished, conceptually on axis I (according to micro-morphological descriptions of thin soil sections) and experimentally on axis III by measuring and interpreting the shrinkage curve (Figure 3.8):

– the inside of primary peds is composed of clay particles and the poral space (variable) termed *micro*. It stays saturated in water with desiccation as long as the water content is higher than that at the air entry point W_B, noticeable on the shrinkage curve (Figure 3.8). The descriptive variables $\overline{Vp}_{mi}$, W_{mi} and $\bar{A}_{mi}$ are microporal volume, micro-water content and micro air content of the pedostructure, respectively. The descriptive variables of the pedostructure are defined in Table 3.4;

– the exterior of primary peds is composed of the surface of primary peds and the inter-aggregate poral space, termed *macro*. The descriptive variables are (Table 3.4): $\overline{Vp}_{ma}$, W_{ma} and $\bar{A}_{ma}$, respectively macro-poral volume, macro-water content and macro-air content of the pedostructure.

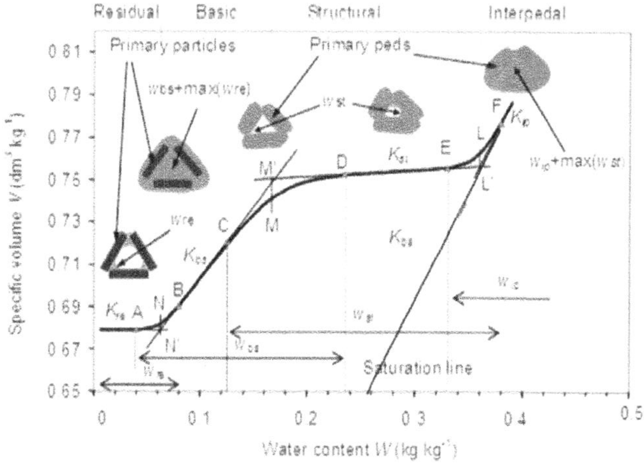

Figure 3.8. *The characteristic shrinkage curve of a soil sample (pedostructure). The different configurations of distribution of water and air in both pore spaces between and within the primary peds of the pedostructure, in relation to the different shrinkage phases of the shrinkage curve (taken from [BRA 04]). For a color version of this figure, please see www.iste.co.uk/braudeau/hydro.zip*

Systems concerned	Pedostructural volumes	Pedostructural water contents	Pedostructural air contents
SREV of the pedostructure	$\bar{V}_{ps} = V_{ps}/M_s$	$W = M_w/M_s$	$\bar{A} = V_{air}/M_s$
Inter-ped poral volume (ma)	$\overline{Vp}_{ma} = Vp_{ma}/M_s$	$W_{ma} = M_{wma}/M_s$	$\bar{A}_{ma} = V_{airma}/M_s$
Intra-primary ped poral volume (mi)	$\overline{Vp}_{mi} = Vp_{mi}/M_s$	$W_{mi} = M_{wmi}/M_s$	$\bar{A}_{mi} = V_{airmi}/M_s$

Table 3.4. *Organization variables of the pedostructure into the micro- and macroporal systems*

3.3.2. *Equations of the hydrostructural equilibrium of the pedostructure*

Recognizing the existence of primary peds in the pedostructure is recognizing two different components of the liquid phase: W_{mi} and W_{ma} submitted to opposite pressures (suction inside primary peds and suction outside of them at their surface). These two components must be featured in

Gibbs' thermodynamic potential equation, written in the form of the Euler equation [SPO 81, BRA 14c]:

$$\bar{G}_\alpha = (\textstyle\sum_\alpha \sum_i \mu_{i\alpha} m_{i\alpha})/M_s, \qquad [3.1]$$

where α represents a phase (solid, liquid or gas) and iα represents the components of this phase. $\bar{G}_\alpha$ is the pedostructural free energy (reported to the mass Ms of the pedostructure) of the phase α (solid, liquid or air) of the pedostructure. The $\bar{G}_\alpha$, like $\mu_{i\alpha}$, are conventionally negative. Consequently, applied to the aqueous solution of the pedostructure, equation [3.1] can be written as:

$$\bar{G}_w = (W_{mi}\mu_{mi} + W_{ma}\mu_{ma}) = \bar{G}_{wmi} + \bar{G}_{wma}, \qquad [3.2]$$

where $\bar{G}_{wmi} = W_{mi}\mu_{mi}$ and $\bar{G}_{wma} = W_{ma}\mu_{ma}$ relate to the two types of water, micro and macro. Note that we cannot write $\bar{G}_w = \mu_w W$, because the relation $\mu_w = \bar{G}_w/W$ is not valid as the potentials cannot be averaged: $\mu_w = \mu_{wma}$ in the macro-aqueous phase and $\mu_w = \mu_{wmi}$ in the micro-aqueous phase.

On the other hand, if h_{mi} and h_{ma} are suctions or retentions of water inside and outside primary aggregates, we necessarily have equality of suctions in the pedostructure in equilibrium at a given water content; and the h^{eq} measured using a tensiometer as follows:

$$h^{eq} = h_{mi} = -\rho_w(\mu_{wmi} - \mu_{wmiSat}) = h_{ma} = -\rho_w(\mu_{wma} - \mu_{wmaSat}) \qquad [3.3]$$

where μ_{wmi} and μ_{wma} are the potentials of the aqueous solution inside (mi) and outside (ma) the internal "micro" system. μ_{wmiSat} and μ_{wmaSat} are the water potential when the entire organization is in a saturated state (no air inside the pedostructure SREV).

As Braudeau *et al.* [BRA 14c] observed, $\bar{G}_w$, $\bar{G}_{wmi}$ and $\bar{G}_{wma}$ (J/kg soil) of equation [3.2] *are constant with the variation in water content*. $\bar{G}_w$ is the sum of both potential energies created by surface charges of clay in the micro- and macroporal spaces. This constancy of $\bar{G}_{wmi}$ *and* $\bar{G}_{wma}$ causes the distribution of water between the micro- and macro spaces such that h_{mi} and h_{ma} remain equal throughout the equilibrium state variation caused by a change in the total water content (e.g. via evaporation).

Let $\bar{E}_{mi} = -\bar{G}_{mi} = -\mu_{wmi}W_{mi}$ and $\bar{E}_{ma} = -\bar{G}_{ma} = -\mu_{wma}W_{ma}$, the specific charge potentials in the solid phase in the micro- and macroporal spaces, respectively; with $\bar{E}_{mi}$ and $\bar{E}_{ma}$ being constant, we deduce from equation [3.3], the following expressions of h_{mi} and h_{ma} given by:

$$h_{mi} = \rho_w \bar{E}_{mi}(1/W_{mi} - 1/W_{miSat}) \text{ and}$$
$$h_{ma} = \rho_w \bar{E}_{ma}(1/W_{ma} - 1/W_{maSat}) \quad [3.4]$$

At equilibrium, the equality $h_{mi} = h_{ma}$ implies the breakdown of W into W_{ma}^{eq} and W_{mi}^{eq}, which are the solutions to the quadratic equation developed from the equality of the above-mentioned equations [3.4]; they are expressed as follows [BRA 14c]:

$$W_{ma}^{eq}(W) = \frac{1}{2}\left(W + \frac{\bar{E}}{A}\right) + \frac{1}{2}\sqrt{\left[\left(W + \frac{\bar{E}}{A}\right)^2 - \left(4\frac{\bar{E}_{ma}}{A}W\right)\right]} \quad [3.5]$$

and

$$W_{mi}^{eq}(W) = W - W_{ma}^{eq} = \frac{1}{2}\left(W - \frac{\bar{E}}{A}\right) - \frac{1}{2}\sqrt{\left[\left(W + \frac{\bar{E}}{A}\right)^2 - \left(4\frac{\bar{E}_{ma}}{A}W\right)\right]}, \quad [3.6]$$

where A is a constant representing the difference in chemical potentials of two types of water in the saturated state:

$$A = -(\mu_{maSat} - \mu_{miSat}) = \frac{\bar{E}_{ma}}{W_{maSat}} - \frac{\bar{E}_{mi}}{W_{miSat}} \quad [3.7]$$

$$\bar{E} = \bar{E}_{mi} + \bar{E}_{ma} \quad [3.8]$$

and W_{miSat} and W_{maSat} are the micro- and macro-water contents at saturation, i.e.:

$$W_{Sat} = W_{miSat} + W_{maSat} \quad [3.9]$$

Equations [3.5] and [3.6] determine the distribution of water in the pedostructure at equilibrium for all water contents W.

3.3.3. *Determining hydrostructural soil parameters*

If a change in water content (evaporation or drainage) is slow enough that the passage from one hydric state (W_1) to another (W_2) proceeds as a sequence of hydrostructural equilibrium states, the shrinkage curves $\bar{V}(W)$ and the water retention curves h (W) are sequential points representing the state of the system (pedostructure) at equilibrium defined and determined by the values of $W_{mi}^{eq}(W)$ and $W_{ma}^{eq}(W)$ given by equations [3.5] and [3.6].

This is actually what is obtained experimentally (sequence of equilibrium states) when the shrinkage curves and retention curves are simultaneously measured using the Typosoil® device (Figure 3.9) on samples (5 cm × 5 cm cylinders of soil) submitted to evaporation, starting from the water saturated state, in the device at 30 or 40 °C [BRA 14c, ASS 14]. TypoSoil® is a recent tool and the only one that can continuously and simultaneously measure the same sample for its both characteristic curves; the retention curve, such that:

$$h(W) = h_{mi}\left(W_{mi}^{eq}\right) = h_{ma}\left(W_{ma}^{eq}\right) \qquad [3.10]$$

whose exact physical equation was given above (equations [3.4]–[3.8]);

– and the shrinkage curve:

$$\bar{V}(W) = \bar{V}\left(W_{mi}^{eq}, W_{ma}^{eq}\right) \qquad [3.11]$$

whose exact theoretical formulation is known as a function of W_{mi}^{eq} and W_{ma}^{eq}, established by Braudeau *et al.* [BRA 14c] (see sections 8.3.2 and 8.3.4).

Figure 3.10 shows an example of both characteristic curves measured using TypoSoil® and modeled according to their theoretical equations based on the thermodynamics of the organized soil medium [BRA 14c]. The result is a perfect superposition of the measured and calculated curves; it initially confirms our view of the pedostructural water as two types of water, micro and macro (or internal and external to the primary aggregates) in thermodynamic equilibrium at each moment of the evaporation process. It also confirms that the correct choice of set of variables used is that defined in the SREV paradigm. The advantage of using the exact theoretical equations of the measured curves is that this adjustment gives an accurate value of the parameters of these equations. These parameters represent then the intrinsic physical properties of the measured and perfectly defined sample (e.g. its microporal water content at saturation W_{miSat}). They are called hydrostructural soil parameters, characteristic of the pedostructure; the methodology used for their accurate determination is explained in Chapters 7 to 9.

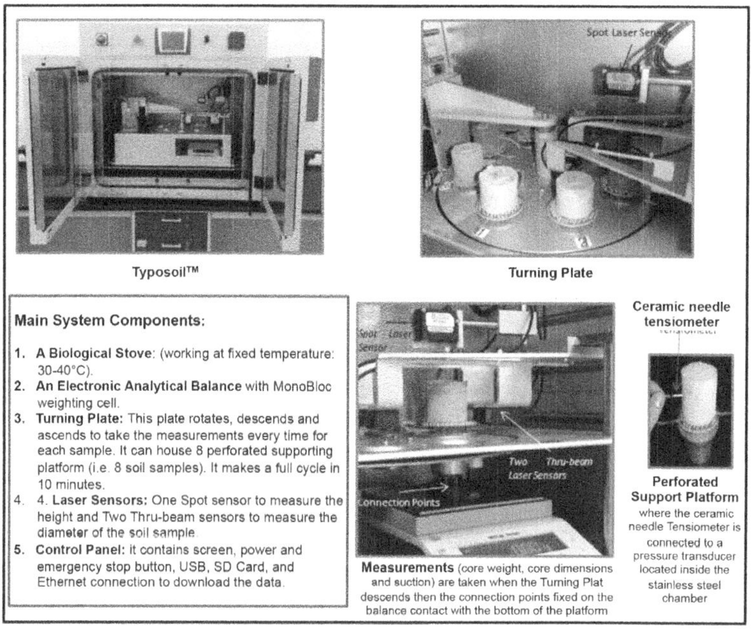

Figure 3.9. *TypoSoil, a device that continuously measures two characteristic curves of soil moisture: the shrinkage curve $\bar{V}(W)$ and the retention curve $h(W)$ [ASS 14]*

3.3.4. *Equations of the hydrodynamic functioning of the pedostructure*

As we have seen above, using the paradigm based on SREV, rather than on REV, allows us to reformulate the thermodynamic equilibrium equation. The same goes for water transfer equations in the soil medium and for their characteristic parameters [BRA 09a, BRA 14a]. There are two types of water transfer to consider in the pedostructure: one is the Darcy flux, usually considered as the only existing flux when the soil is described according to the REV principle as a continuous porous medium having no interaction with water. The second is the flux of water exchanged locally between two porous systems, micro and macro, to rebalance the pressures of the two systems ($h_{mi} = h_{ma}$), following a variation in the macro-water content W_{ma}.

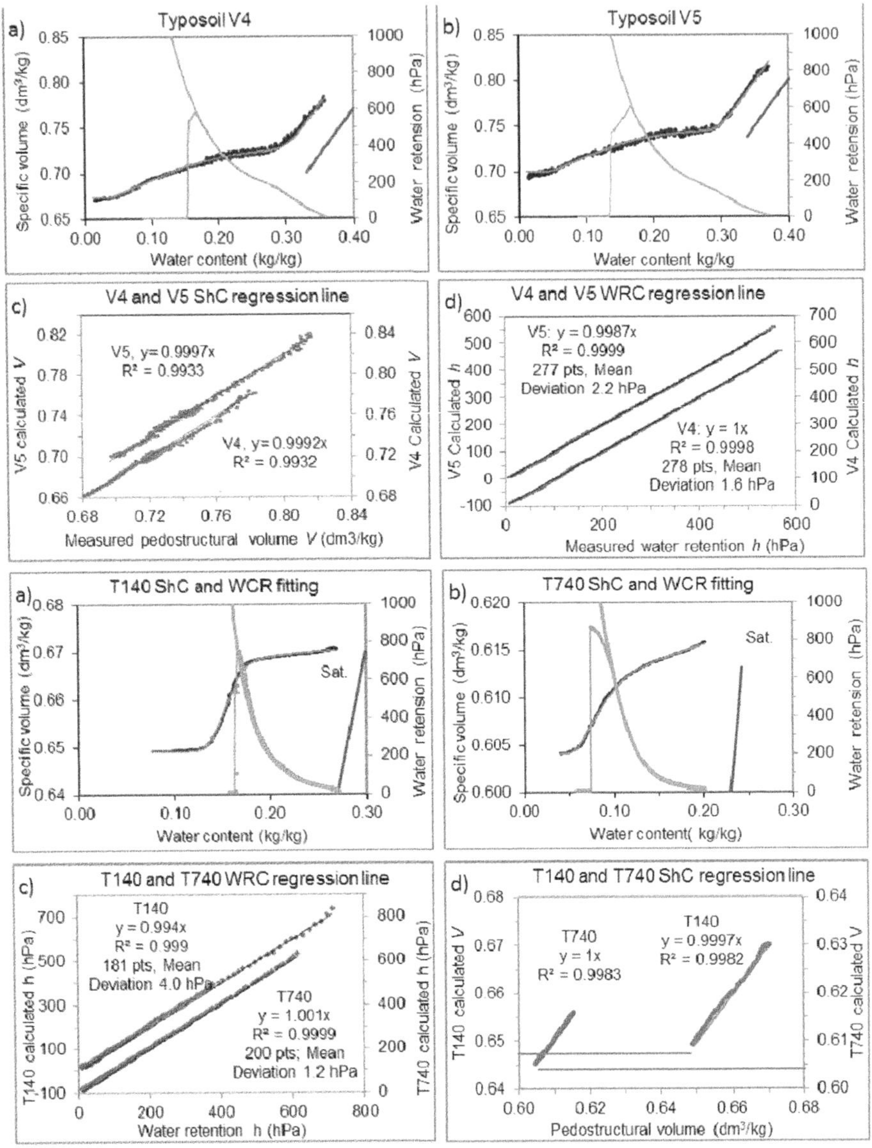

Figure 3.10. *TypoSoil measurement results of two distributions of the same silt-clay soil and two different ferralitic soils with regard to the clay content: a) and b): measured shrinkage curves (in blue) and the theoretical shrinkage curves (in red), and the measured retention curves (in yellow) and the theoretical retention curves (in green); c) and d) statistical comparison of the measured and the calculated shrinkage curve (ShC) and the retention curve (WRC) (taken from [BRA 14c])*

a) Considering that the Darcy flux is the flow of water between aggregates (W_{ma}) and with a variation in W_{ma}, the micro-water content W_{mi} of primary peds increases or decreases according to this variation, to maintain local equilibrium of the retention pressures between two porous systems
(h_{mi} = h_{ma}), the Richards equation for the transfer of water in the pedostructure can be written as:

$$dW/dt = \rho_w \bar{V}\, d\left[k_{ps}(dh/dz + 1)\right]/dz, \qquad [3.12]$$

where k_{ps} is the hydric conductivity of the pedostructure, z is the elevation (positive upwards) and dh/dz is the pressure gradient (Braudeau and Mohtar [BRA 09a, BRA 14a, BRA 14b]).

In fact, k_{ps} will only be a function of W_{ma}, the water in the macroporosity because the micro- and macro-water exchange is local, and the amount W_{mi} at time t remains constant over the time dt in the pedostructure element between the surfaces z and z+1. This gives the relation:

$$dW/dt = dW_{maDarcy}/dt + dW_{maLocal}/dt + dW_{mi}/dt = dW_{maDarcy}/dt, \qquad [3.13]$$

where $dW_{maDarcy}$ is the variation in the macro-water content of the pedostructure element due to the Darcy flux, whereas $dW_{maLocal} = -dW_{mi}$ is the variation in W_{ma} due only to the local rebalancing of pressures between the micro- and macro-aqueous phases of the pedostructure. We therefore develop the Richards equation [3.12] as a function of W_{ma}, as follows:

$$\frac{dW}{dt} = -\rho_w \bar{V}\left\{\frac{\Delta k_{ps}}{\Delta W_{ma}}\left(\frac{\Delta h-\Delta z}{\Delta z}\right)\frac{\Delta W_{ma}}{\Delta z} + k_{ps}\frac{\Delta(dh/dz)}{\Delta z}\right\}, \qquad [3.14]$$

where Δh, Δz , Δk_{ps} and ΔW_{ma} represent the differences in the values concerned between the two close soil levels (see Figure 3.11).

However, as k_{ps} is completely defined by equations [3.12] and [3.14], we can use the HYPROP® measurement device (UMS GmbH, Munich, Germany), shown in Figure 3.11, or any other equivalent tool (section 8.3.5) that provides the exact experimental conditions corresponding to these equations.

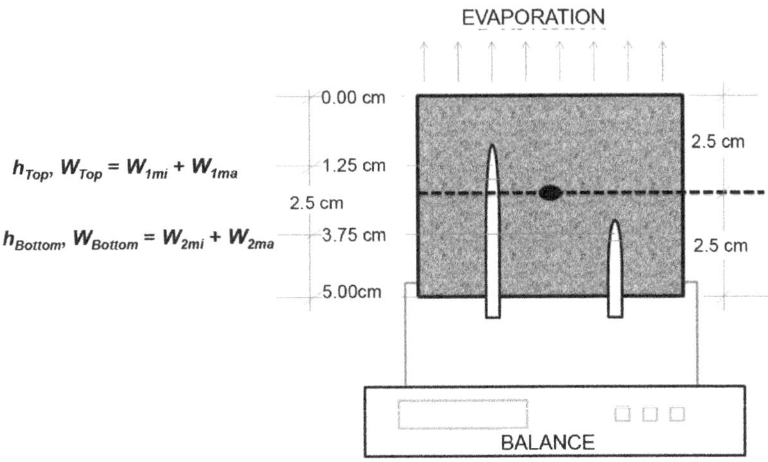

Figure 3.11. *Diagram showing the HYPROP® principle used to measure the hydraulic conductivity of a cylindrical soil sample (250 cm² by 5 cm in height). The water content decreases due to evaporation at the top of the cylinder, left empty. Continuously measuring the total weight of the sample and suction (h_{top} and h_{Bottom}) at two depths, using 2 mini-tensiometers, allows us to calculate W_{mi} and W_{ma} at these two levels at each moment, and hence the variation in the hydraulic conductivity as expressed in the text*

We shall see in part two of this book (Chapters 8 and 9), however, that these can only be physically exploited without any assumption or approximation as currently required [SCH 10], if the soil sample has been previously characterized by measuring the characteristic moisture curves and their determining hydrostructural parameters. In fact, using these parameters and the unique theoretical curves linking h to the water contents W_{mi}^{eq} or W_{ma}^{eq} (from the sample analyzed in equation [3.4]), these water contents can be calculated for each value of h measured *locally* at two heights z_1 and z_2 of the sample, by the mini-tensiometers of the measurement device (Figure 3.11).

All terms of equation [3.14] can therefore be calculated, as we will see in Chapter 8 (3.5), allowing us to see that k_{ps} is an exponential of W_{ma} in the form:

$$k_{ps} = k_{ps}^{0} W_{ma} Exp\left(\alpha_{ps} W_{ma}\right). \quad [3.15]$$

The constants α_{ps} and k_{ps}^0 are determined by processing the measurement data in a particular way using Excel, which will be shown in Chapter 9.

Note that equation [3.15], the exact physical equation of k_{ps} found experimentally, could not have been defined without the definition and the systemic formulation of the hydric conductivity of the pedostructure by equations [3.12] and [3.14], systemic because their variables are all systemic, nor without the thermodynamic formulation of the soil water retention pressure according to the two types of water W_{mi} and W_{ma} at equilibrium (equations [3.4] to [3.10]).

In fact, before moving onto the K_{ps} equation [3.15], the fundamental and surprising result initially found from the measurement results is that the term $\left(\frac{\Delta h-\Delta z}{\Delta z}\right)\frac{\Delta W_{ma}}{\Delta z}$, factor of dk_{ps}/dW_{ma} in the Richards equation [3.14], is a simple exponential of W_{ma}, with a correlation coefficient greater than 0.999 (see section 8.3.5, Figure 8.8).

b) Now considering the second type of local water transfer "micro–macro" in the pedostructure element and taking into account equation [3.13], the corresponding transfer equation is:

$$dW_{mi}/dt = k_{mi}(h_{mi} - h_{ma}), \qquad [3.16]$$

where k_{mi} is the characteristic transfer coefficient of microporosity. From wider research into the consistency of this parameter, it was considered fixed, i.e. independent of the water content of the pedostructure or that of primary peds [BRA 06b, BRA 09a]. It is determined by measuring the variation in the height of a bed of fine aggregates (fine earth sieved through 2 mm) submersed in water in their dry state. The rate at which the height varies indicates the water absorption rate. The microporosity decreases exponentially according to a theoretical equation established in the pedostructure paradigm [BRA 06b]. The adjustment of the theoretical curve to the measured curve (height of the aggregate bed with time just after the immersion) provides the parameter k_{mi}, that of equation [3.16].

In summary, the *hydrostructural parameters of the pedostructure* define the states of the thermodynamic equilibrium of the pedostructure according to the water content, and together with the *hydrodynamic parameters* just seen (k_{ps0}, α_{ps} and k_{mi}), form what are called *pedostructural parameters*.

These parameters define and represent the hydrostructural and hydrodynamic functioning of the pedostructure. The question now becomes: how is the pedostructure, a system whose functioning is entirely described by pedohydric parameters, integrated into the overall functioning of the organized system of the pedon and its horizons?

3.3.5. *The Kamel® model of the hydrostructural functioning of a pedon*

After developing this new physics of the organized soil medium, the Kamel® model was implemented [BRA 14a] not only to take into account thermodynamic equations of water in the newly established pedostructure [BRA 14c, BRA 14d], but also to take into account the sub-systems present and connected to the pedostructure in the same soil horizon (volumes occupied by roots, biological macropores, coarse elements, etc.). Through its implementation, true interdisciplinary relationships could be considered between the Kamel® model and models from other agro-environmental disciplines whose subject of study is dependent on soil.

3.3.5.1. *Replacement with new equations*

The equations to be replaced are already systemic (variable and systemic parameters) but are not derived from thermodynamic laws, as are the new equations. With regard to the shrinkage and retention curves of the pedostructure, the number of parameters used remains constant, and we know how to move from one set of parameters to another, thus having no need to replace the old equations from these two curves with new ones from the model. However, the new equation of the hydric conductivity curve $k_{ps}(W)$ only has two parameters rather than the previous four. This is due to the fact that the new thermodynamic equation of the retention curve $h^{eq}(W)$, equations [3]–[6], is valid over all water contents and not limited to only the validity zone of the tensiometer [BRA 14c, BRA 14d].

3.3.5.2. *Considering other sub-systems associated with the pedostructure in the soil horizon*

In the original version of the Kamel® model, the soil horizons were only composed of the pedostructure ($V_{ps}=V_{hor}$), and therefore parameters of the pedostructure were sufficient to characterize the hydrostructural functioning of the horizon. As we shall see, these parameters of the pedostructure are still valid in the new version, but must be completed by parameters relating to the properties of other sub-systems that share the soil horizon with the pedostructure, as shown in Table 3.5 (taken from Braudeau and Mohtar [BRA 14a]).

SREVs of concern	Internal components	Morphological parameters	Functional parameters
– Pedon	– Surface layer – Horizons	S_{xy}, L_z,	Bottom conditions Surface conditions
– Surface layer	– Pedostructure, clods – Macro-inter aggregate space	H_{surf},	$K_{satSurf}$, $V_{surfSat}$, a_{surf}, b_{surf}, c_{surf}, d_{surf},
– Horizons	– Succession of SRELS	H_{hor}. Zi,	SREL parameters
– SRELs	Pedostructure V_{ps}, macropore volumes (Vp_{bio}, Vp_{fiss}, …) and solid elements (V_{stones}, V_{roots})	H_{iSat}, S_{xy},	– Field saturated hydraulic conductivity K_{sat} – Volumetric % of components a, b, c, d … in the horizon – Pedostructureparameters

Table 3.5. *List of the hydrofunctional subsystems of a pedon SREV, internal components and corresponding parameters. Parameters are explained in the text*

The parameters listed in Table 3.5 are considered as fixed data of the organized and functional soil environment. However, we know that the activity of the biological systems present in the horizon (root growth, release of organic substances by roots, activity of microfauna, etc.) modify some characteristics over time, during a crop cycle for example. It is not difficult to take into account this slow variation of parameters over time in the model.

Braudeau and Mohtar [BRA 14a] have provided the methodological guidelines to consider other organizational volumes usually present in a soil horizon: biological macropores, plant roots, coarse mineral elements, such as stones, rocks, nodules, etc. These methodological guidelines not only concern the writing of complimentary equations in Kamel®, which puts the activity of the pedostructure into perspective relative to new structural volumes added to the horizon, but also concern the method used to estimate additional parameters to be observed or measured at the pedon scale in the field. The best example is measuring the hydric conductivity at saturation (K_{sat}), traditionally done in the field using a disk or double ring infiltrometer [ANG 00]. K_{sat} is the sum of the hydraulic conductivity of the pedostructure and (k_{psSat}) is that of the macroscopic porosity of biological origin beyond the pedostructure, and unlike that of the pedostructure, is highly spatially variable. This spatial variability does not, therefore, depend on the texture but on other factors associated with the macro-biological activity of soil. These must be known if we wish to use pedotransfer functions instead of taking measurements to estimate K_{sat}.

3.3.5.3. *Discretization of horizons in SRELs (Structural Representative Elementary Layers)*

Simulating the transfer of gravimetric water and the associated processes inside and across the pedon requires the discretization of the environment into representative elementary layers of the structure of horizons. Assuming that the soil horizon has homogeneous hydrostructural properties, and that there is an SREV of the corresponding horizon in the soil mapping unit, then each soil horizon can be split into thin horizontal layers that have physical properties (parameters of the pedostructure) of the horizon considered, and that are at different equilibrium states according to the water content of the soil (Figure 3.12). These layers are called structural representative elementary layers (SRELs). They have a minimum thickness of 2 cm in Kamel®. Furthermore, a pedon of 1 m width is considered sufficiently wide to represent a soil mapping unit insofar as certain pedological characteristics (fissures, stones, etc.) are observable at the pedon scale in the field.

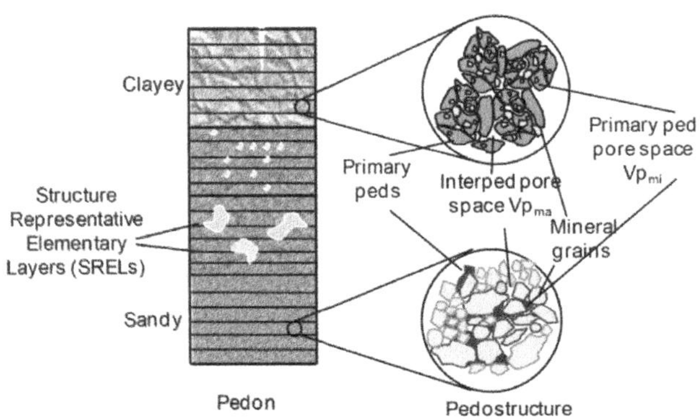

Figure 3.12. *Diagram of the pedon and its internal organization as modeled by Kamel®. The separation of horizons into SRELs with a width of 2 cm allows the physical and the systemic modeling (non-empirical) of thermodynamic equilibrium states of each SREL, as well as the dynamics and distribution of water types in each part of the pedon*

The descriptive variables of an SREL include those of the pedostructure as well as other elements, such as roots, biogenous macropores, stones, etc., whose volumes are related to the mass of the total pedostructure of the layer (M_{psL}). Table 3.5 shows the different hydrofunctional sub-systems of the pedon, such as SRELs, whose variables and parameters must be defined according to the SREV concept in order to be compatible with other organization levels of the pedon (horizons and pedon).

In particular, the mass of the pedostructure included in the SREL is useful as a reference rather than the total mass of solids in the layer. This allows us to keep the variables and properties of the pedostructure as measured among the additional variables of the new version of the SREL.

Let us assume, for example, that the SREL of a soil horizon is composed of a certain volume of the pedostructure and other organizational elements, such as volumes of stones, roots, macropores of biological origin, etc. All of these volumes, including that of the pedostructure, should be estimated as fractions of the total volume of the representative horizon layer, as shown in equations [3.17]–[3.20] below. However, only the mass of the pedostructure contained in the SREL, excluding other solid masses, will be used as the reference mass for different extensive variables of the SREL. This is why the discretization of the horizon into 2 cm wide SRELs should be done when the horizon is in a saturated state (or at the same hydraulic state throughout the whole horizon): so that the mass of the pedostructure can be considered uniformly distributed horizontally and vertically throughout the horizon. Each SREL of the horizon will therefore have exactly the same characteristic parameters as those of the horizon.

Let $a = \left(V_{ps} + V_{fiss}\right)/V_{horizon}$ be the volumetric proportion of the pedostructure and associated cracks (due to shrinkage–swelling) in an SREV of the horizon considered. Let $\bar{V}_{ps} = V_{ps}/M_{ps}$ (dm^3/ kg), the specific volume of the pedostructure measured in the laboratory using a soil sample representing the pedostructure of this horizon. Thus, the mass of the pedostructure in the discretized volume $\left(V_{lay}\right)$ of the layer SREL of the horizon will be:

$$M_{psL} = aV_{layer}/\bar{V}_{ps}, \qquad [3.17]$$

where M_{psL} is the overall mass of the pedostructure in the layer, and $\bar{V}_{ps}$ is the specific volume of the pedostructure. Consequently, if b is the volumetric fraction of stones in the horizon, c is the volume fraction of biological spaces, d is the proportion of roots, etc., the specific volume $\bar{V}_{layer}$ of an SREL of the horizon will be:

$$\bar{V}_{layer} = \bar{V}_{ps} + Vp_{fiss}/M_{psL} + V_{stone}/M_{psL} + Vp_{bio}/M_{psL} + V_{roots}/M_{psL} \quad [3.18]$$

$$\bar{V}_{layer} = \bar{V}_{ps} + \overline{Vp}_{fiss} + \bar{V}_{stone} + \overline{Vp}_{bio} + \bar{V}_{roots} \quad [3.19]$$

and the specific volumes of different sub-systems of an SREL are as follows:

$$\left(\bar{V}_{ps} + \overline{Vp}_{fiss}\right) = a\,\bar{V}_{layer};\ \bar{V}_{stone} = b\,\bar{V}_{lay};\ \overline{Vp}_{bio} = c\bar{V}_{layer};\ \text{and } \bar{V}_{roots} = d\bar{V}_{layer} \quad [3.20]$$

Equation [3.20] defines the coefficients a, b, c and d as the characteristic parameters of a horizon of the pedon, which must be estimated by the simple observation of the soil profile *in situ*. The sum of these coefficients is equal to 1. Similar to how we distinguished the porous volumes $\overline{Vp}_{fiss}$ and $\overline{Vp}_{bio}$ of the layer as well as the porous volumes $\overline{Vp}_{mi}$ and $\overline{Vp}_{ma}$ of the pedostructure, we also distinguish the water contents in these volumes, i.e. W_{fiss}, W_{bio} and W_{ps}, respectively. Note that of all the specific volumes considered in equations [3.18]–[3.20], $\bar{V}_{ps}$ and $\overline{Vp}_{fiss}$ are the only properties of the pedostructure that vary with the hydric state of the soil. They are the functions of the pedostructural water content W_{ps} and are determined according to the parameters of the shrinkage curve.

3.4. Systemic mapping of soil in the landscape

How do we use pedological maps as a geo-referenced *systemic* representation of primary soil units, characterized by their representative pedon?

3.4.1. *Hierarchical hydrofunctional mapping units of the landscape*

A typical pedological map is a cartographic representation of the space occupied by soil organizations, as defined, described and characterized on axes I and II. As we have seen, axis III has been overlooked in pedological mapping due to the incompatibility of the REV and SREV approaches. Therefore, the typical pedological map is not completely systemic in the sense defined. However, it is possible to come close to this definition by taking this map and performing cartographic delimitation at several levels of nested organizations (units of relief including formation units, which include primary soil units) according to the hierarchical structuration diagram of natural organizations, as shown in Figure 3.4. Note that the primary soil units are classed as 'primary' because they are represented by a representative pedon which, as we have seen, is characterized and modeled systemically by the Kamel® model.

The relief units and geomorphological formation units are easily recognized and can be sectioned into nested systems by interpreting satellite and/or aerial photos. The last level, soil mapping units within geomorphological units, requires a field survey in order to define "primary soil units" i.e. *homogeneous zones with regard to the hydrostructural behavior of horizons*. These soil units are usually limited in number (up to a maximum of 3) in the same geomorphological unit. Each unit will be represented by its representative pedon, which in fact is an SREV of the soil mapping unit.

By associating the Kamel® model of the representative pedon with the mapping of the three nested levels just presented, we obtain a complete systemic description of soils in a zone or region on the three referenced axes. The digitization of these mapping units, associated with a georeferenced database containing the modeling parameters (by Kamel®) of all representative pedons of the zone, leads to what we call the soil spatial reference information system (SIRS-Soils) of a given zone.

An example of soil maps resulting from systemic mapping [BRA 01] is given above. The pedological and physiographic maps, extracts of which are given in Figure 3.13, are graphical outputs (layout of Arc View projects) of the SIRS-Soils of the irrigated plot of Cebala. Both extracts represent the same portion of the map. We find that the units are nested: the numbers in the units of formation refer to matter (nature and implemented), and the letters in the soil units indicate textural class.

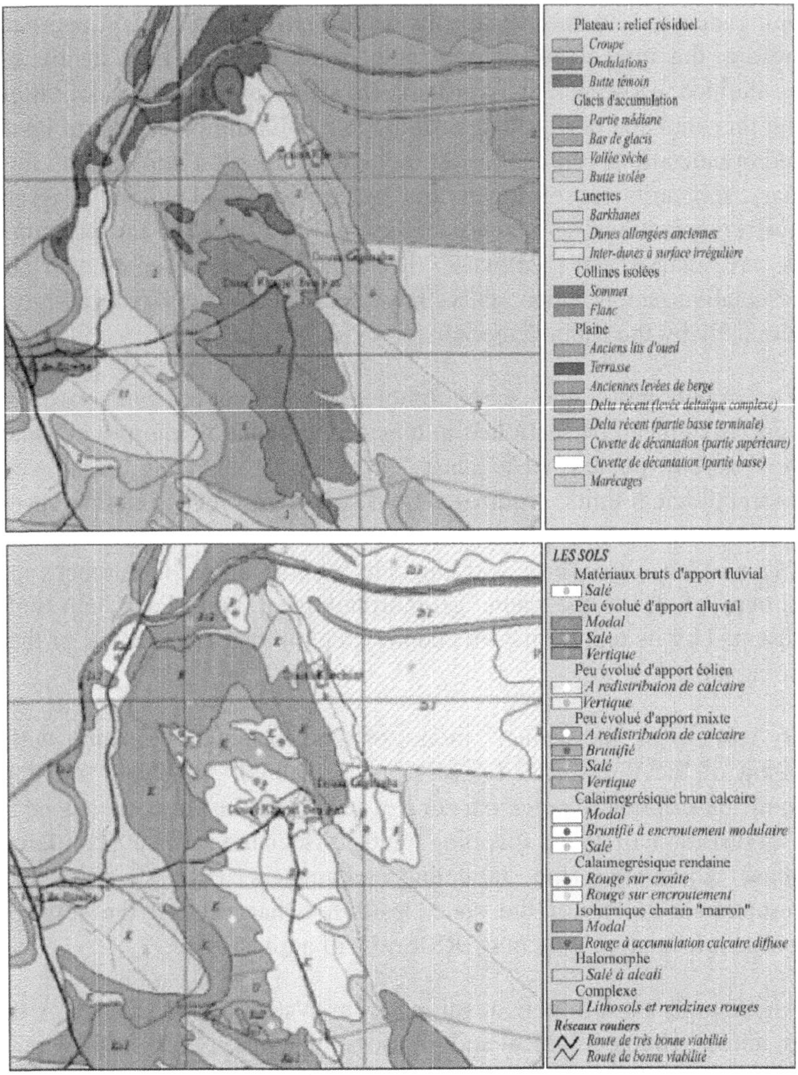

Figure 3.13. *Reduced extracts a) of the soil map 1/20,000 of the irrigated perimeter in Cebala and b) of the corresponding physiographic map, with their respective legends [DER 01]. For a color version of this figure, please see www.iste.co.uk/braudeau/hydro.zip*

The characterization of a primary soil unit of the soil map (see Figure 3.13(a)) will be considered equal, on average, to the characterization of the hydrostructural functioning of its representative pedon. This, composed of its different horizons, can be comprehensively modeled by the improved version of the Kamel® model recently described by Braudeau and Mohtar [BRA 14a] and which satisfies the systemic and *thermodynamic* approaches defined above.

The characteristic soil parameters required for this Kamel® modeling are those of the physical equations established according to this approach for three organization levels of the pedon summarized in Table 3.5; these are:

1) hydrostructural and hydrodynamic pedostructure parameters of each horizon of the representative pedon, i.e.:

i) parameters of the hydrostructural equilibrium equations of the pedostructure, parameters of theoretical equations of the water retention curve ($h\ (W)$) and the shrinkage curve ($V\ (W)$);

ii) parameters of the water transfer equation in the pedostructure (k_{ps} (W) and k_{mi}).

2) parameters of organization and hydric conductivity of the horizons at saturation, from the morphological observation of soil horizons in a pedological pit; that is to say:

i) the descriptive parameters in terms of volume percentages of the horizon composition occupied by the pedostructure (fine earth), the root system, biological macroporosity, pebbles or other coarse elements, and the number of vertical fissures;

ii) the parameter k_{sat} to be measured at each horizon using conventional tools.

3) organizational parameters of the pedon from a description of the soil surface and representative soil profile, such as:

i) some characteristics of the soil surface (vegetation cover, surface crust, fissures, microrelief) dealt with by the Kamel® model;

ii) the average depths of the boundaries of each horizon.

3.4.2. *The SIRS-Soils*

The SIRS-Soils of a given zone provides information, such as the limits of soil units defined systemically in a hierarchical structure (relief, formation and soils). Included in this structure, in the first organization level, they are defined with minimal error according to the three description axes of the systemic approach in Figure 3.6. The novelty, compared with the soil mapping of the 1970s–1980s that studied soil organization on two "organizational" axes and genesis, is the hydrostructural modeling of the pedon on axis III. This is why the current pedogenetic classification by naturalists [CPC 67] is still justified and must be retained for naming soil units at the geomorphological level: it will be completed by a typology of hydrostructural properties of soils in primary soil units.

The two inherent problems in soil science presented above are illustrated in Figures 2.1 and 2.2, and have been solved with the launch of the SIRS-Soils. We went from an empirical characterization and modeling of the "soil medium" to characterization and modeling linked to the physical theory of the organization of this "soil medium", the systemic and thermodynamic physics of the interaction between the soil and the water and the soil structure.

The new challenge for soil science is the generalization of the methodology to develop the SIRS-Soils with many existing pedological databases. This relies on the implementation of these new physical theories and helps characterize and model the hydric and structural functioning of the pedon (Figure 3.12) representative of one soil mapping unit. For the first time, this will provide a means of identification and comparison of soils according to their hydrostructural functioning, defined in connection with their internal organization (axis I) and pedogenesis (axis II).

4

The General System: General Model of Scientific Disciplines Related to the Study and Management of Natural Areas

4.1. The human system, system of study or management of a natural area, isomorphic to the general system

As mentioned in Chapter 3, we will have another look at the representation that Le Moigne gave of the general system (GS) of level 9 (Figure 3.2) to ideally represent the relationship between man and a particular object in the natural environment, the soil, whether for research, management or use. We, therefore, consider that Le Moigne's GS model [LEM 94], provided that it is associated with the systemic description defined above, is the universal model of the organizational system implemented by man to take charge of the knowledge, use or management of organized natural spaces with soil being the supporting medium.

Figure 4.1 represents the isomorphism of the organization between the GS and the human system (HS) ideally organized for the study or management of natural spaces (organized or not) whose soil forms the primary infrastructure. The three sub-systems necessary for the scientific discipline (the HS) represented are: the guarantors of the discipline, the spatial reference information system (IS) and the natural area concerned.

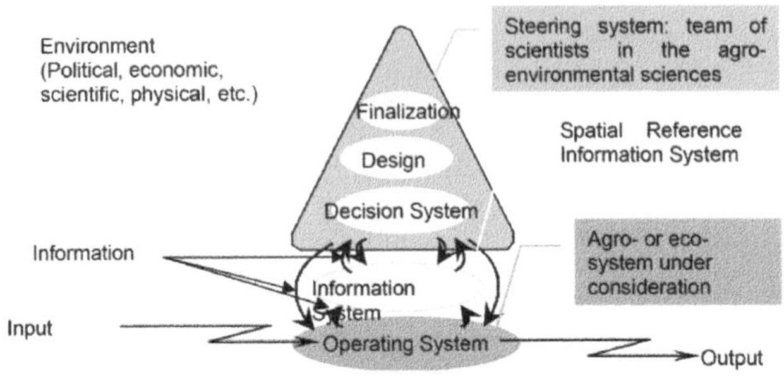

Figure 4.1. *General system model (adapted from [LE 94]) as a model showing the ideal relationship between man and the object of study, research or use*

This diagram of the GS highlights the importance of the IS for the proper management of the operating system (OS). The latter is materially in contact with the external environment due to flows entering and exiting the system, which pass through and interact or not with it.

These flows are controlled at the entrance and exit by the steering system (SS), based on information and knowledge of the operating system stored and organized in the IS. This is an ecological and agronomic modeling platform of recognized soil mapping units of the OS, whose activity (consumption, production) and structural evolution (cyclical variation, aggradation, degradation) we will simulate under the influence of inputs and flows of matter (in particular water), which pass through and interact. The IS, therefore, forms the knowledge base necessary to optimize the control of resources (inputs and OS).

However, we must highlight the need to introduce a framework of systemic physics (as defined above with reference to the three graduated axes) into Le Moigne's GS model [LEM 94]. Le Moigne never mentioned the possibility that this representation of level 9 of the GS could be considered as a universal model, for the relationship between man and the object of study, use or management. It was impossible for Le Moigne to consider this possibility because for that he would have had to make a radical distinction between the OS, representing the systemics of a natural organization and the GS or even the SS, systemic representatives of a human organization as previously defined. In fact, saying this in his new discourse of the method [LEM 94, p. 43]: "The pertinence precept: that any object that we consider is defined according to the implicit or explicit intentions of the modeler", he could not appreciate in the OS, which is the systemic representation of the natural organization being studied, a totally objective description as Descartes stated: "My thoughts do not ascribe any necessity to things" (cited by Le Moigne about the causality precept, p. 37). On the contrary, he refuted this possibility of objectivity, only paying attention to the "behavior-purpose visible to the senses but still subject to the intentions of the modeler." He stated: "In the cause-effect explanation, intelligence therefore substitutes, with a productive generalization, the interpretation (or understanding) of behavior-purpose", thereby dodging the issue with the structure-behavior relationship (in his opinion unknowable), which however formed part of the systemic paradigm forming the basis of his description methodology by triangulation shown in Figure 3.1.

We have, therefore, not followed Le Moigne in this approach, by reformulating the principles of systemic description in the universal model graduated defined above (see section 3.2.2: constituents and form of organizations; evolution of these organizations, internal-external activity), we have kept the reality of this important challenge which is the "systemization" of the natural organization (like the soil): a prerequisite for its study or use by man according to the GS model.

4.2. Natural systems, OSs of the GS

As we have seen above (Figure 4.1), the three sub-systems: OS, IS, SS of the GS deserve to be called a system if they can be described on three axes of the systemic space of description. Their description on the three graduated axes: 1) makes them as systems and 2) put them in correspondence relationships according to the three axes. The combination of these three sub-systems, each open to its environment of the same nature (as shown in Figure 4.2), "makes" the GS.

In the fields of environmental or natural sciences, we know that man plays no part in the creation of natural organizations, whatever they are, but has complete responsibility for a system that he has created and implemented; and he can describe, in detail, its composition, structure and capacity for action. Man also plays a role in the management, use, research, etc. of his natural organizations that he simply sampled or defined to create *natural systems* (NS), which are then transformed and used depending on his needs. It is precisely this *system of study* or *management*, implemented to study or use this *NS*, which can be built according to the *GS model* (Figure 4.1) and can be called a HS. While the NS, existing only after the *conscious delimitation by man of a volume belonging to the natural organization, or of a volume made of natural elements*, can only be the OS of the GS. It cannot, in any case, be identified as a GS, contrary to what may be understood from Le Moigne's presentation of the GS theory (see the new charter of the discourse on method, Table 3.1).

In this sense, a *natural system* is defined as being an OS of a GS (with all HSs being isomorphic); it comes from the spatial boundary setting (zonation, segregation or sampling) of a natural organization, at a given scale. Therefore, the definition of the NS is clearly established based on Le Moigne's GS model [LEM 94] plus the principles of the systemic description of a natural organization on the three reference axes. It is consistent with the methodology of the systemic description of the OS (identified with the NS) leading to a systemic mapping of the studied organization which will be stored in the IS of the corresponding GS.

4.3. Information systems of human systems implemented for the study or management of natural systems

According to our view of the HS, isomorphic to the GS, we can define several categories of HS, depending on the perspective with which it was created. This can be a system of management, use or maintenance of the OS, if the OS is to be used, managed or maintained; it can also be a scientific research system if the OS represents the natural organization to be studied, understood and modeled. Both orientations of the GS are represented in the example given in Figure 4.1.

Note that, in the figure, the IS of the GS is analogous to our current geographic information systems (GIS), apart from the fact that they are not completely systemic: the mapping units, natural or not, of current GIS have not been developed on the plane of axes I and II in relation to the hydrofunctional axis (axis III) of the systemic reference axes as with the spatial reference information system of soils (SIRS-Soils) mentioned in section 3.4.

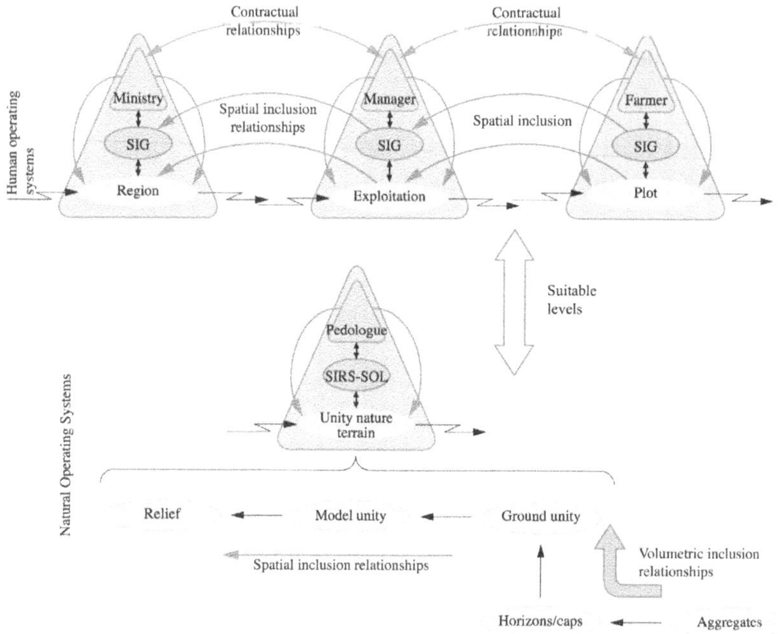

Figure 4.2. *Operating systems of the natural and human environments; giving rise to, respectively, the SIRS-Soils and the GIS. The match between the different SG (scientific disciplines, managing systems etc.) is between their respective Information Systems [BRA 01]. For a color version of this figure, please see www.iste.co.uk/braudeau/hydro.zip*

Whatever the category of HS (study, research, management or use), the SS will control and act according to the purpose of the HS (purpose that determines the category). The IS is only linked to the category as it must contain, at minimum, the information required to steer the OS; without being restricted to this minimum requirement related to the specificity of the OS: due to the compatibility of information between the organization levels in the systemic approach, the IS can store all possible knowledge about the OS at its different organizational levels.

Another type of limitation of the IS is when the HS is considered as the OS of a higher level HS, its freedom of action will be reduced by the orders of the higher level HS, which will act according to its own IS. An example of this HS hierarchy is given in Figure 4.2. We must guarantee the compatibility of IS of two nested HS; which the systemic approach can allow [BRA 01]. Note that one category of HS theoretically avoids any limitation of its IS; this is the "academic discipline" that is developed according to the GS model whose main objective is the understanding of the natural object, independently from any notion of use.

4.4. Hydrostructural pedology and its own spatial reference information system: the SIRS-Soils

This leads to a more accurate understanding of the GS model presented in Figure 4.1 and its close relationship with the systemic approach as we have defined here as the application of the systemic description on the three reference axes (see section 3.2) to the objects that have to be considered as systems (OS, IS, SS, GS, NS, HS). Thus, the systemic approach applied to the scientific discipline in charge of the study of soils, will agree that:

1) The GS, as represented in Figure 4.1, is the universal model of a scientific discipline whose object of study or investigation, the OS, is virtually reproduced (modeled) and stored in an IS conforming to the three description axes of the systemic paradigm. With the OS being a defined space composed of soil distributed in this space, the IS will inevitably be a SIRS: Spatial Reference Information System. It will be based on the systemic concepts explained above for soils (pedon, horizons, SREV, SREL, pedostructure, primary aggregates) to define and delineate its mapping units. The corresponding scientific discipline will be called "Hydrostructural Pedology" in reference to the hydrostructural properties of soil that this discipline covers, as we will see later on.

2) The SIRS is an organized memory, a database that, when complete, includes all information specific to the OS and its sub-systems, their variables and organizational parameters and functioning. The purely empirical data, undefined on one of the three description axes, has been gradually abandoned with the perfection of the system. This SIRS will be called the SIRS-Soils if it represents the pedological cover of the study area [BRA 01].

3) The OS studied is the NS, a piece of the natural organization that has all the qualities of an organized system: 1) it is defined in its external form by a boundary preventing all transfer of solid matter belonging to the structure but permeable to liquid and gas phases and 2) it is composed of an assembly of systems, whether organized or not, but with accurate, well-defined boundaries; all boundaries are defined on axis I.

Thus, the scientific discipline in charge of the study of soils, organized according to the GS model, and satisfying the systemic approach criteria, is typically the hydrostructural pedology defined above as with its specific information system: the SIRS-Soils. The latter, already presented in section 3.4, is the repository of the description and systemic modeling of soils in a study area (the NS); it constitutes heart of the discipline shown in Figure 4.3, isomorphic to the GS (Figure 4.1) of the universal relationship between man and its object of study, management or use.

Hydrostructural pedology is a new discipline in the agro-environmental sciences created due to the connection of axis III, axis of the modeling of the hydrostructural properties of the pedon, with the plane of axes I and II of the classical pedology: qualitative description of the internal organization of soils, of their spatial distribution and their genesis. It should gradually be confirmed within the agro-environmental sciences with recognition from other disciplines of new concepts highlighted in red in Figure 4.3. They are mainly three in number, coming from the application of the GS concept and the systemic description repository concept, to the description, characterization and modeling of the natural environment: i) the specific laboratory for the discipline, ii) the SIRS-Soils, which will be described later on, and iii) the Kamel® model, which entirely represents the internal organization and the hydrostructural functioning of the representative pedon of the soil mapping unit.

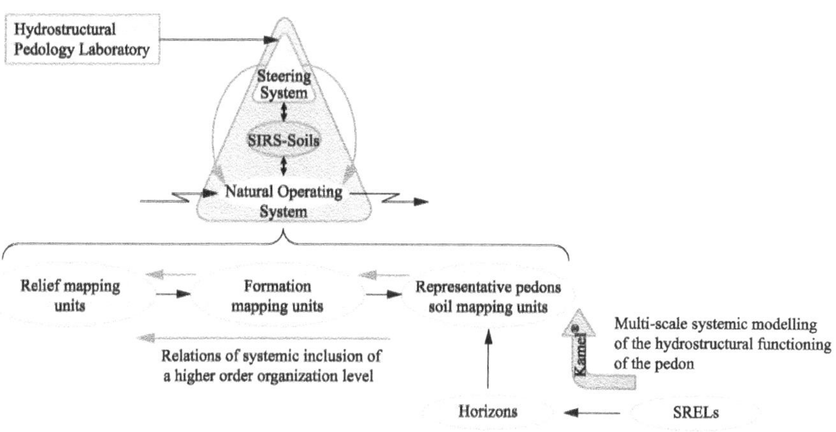

Figure 4.3. *Schematic representation of hydrostructural pedology, isomorphic scientific discipline of the general system composed of its three fundamental sub-systems (SS, IS and OS), respectively: the hydrostructural pedology laboratory, the SIRS-Soils and the natural system of study whose representative pedons are modeled by the systemic soil-water model: the Kamel® model. For a color version of this figure, please seewww.iste.co.uk/braudeau/hydro.zip*

The SIRS-Soils is a generic product of the new discipline; it is valid for all soil situations: it will always be composed of a map of mapping units of three nested organizational levels (Figure 4.2): that of primary soil units, of geomorphological units, and units of relief, together with modeling parameters of the pedon, representative of each primary soil mapping unit.

5

Emergence of a New Scientific Discipline: Hydrostructural Pedology

5.1. Where hydrostructural pedology fits into the natural sciences

Figure 5.1 shows where hydrostructural pedology fits in the agro-environmental sciences, next to hydropedology, a recent discipline, which arose following the combination of two traditional disciplines: pedology and surface hydrology to solve agricultural and environmental issues [LIN 03, LIN 12, LIN 06].

Hydrostructural pedology plays a role that stayed unoccupied in the agro-environmental sciences for lack of scientific theory explaining the hydrostructural and thermodynamic functioning of its soil environment. This role involves "the characterization, mapping and physical and systemic modeling of soil hydrostructural properties", that is to say, hydrostructural properties the structured physical medium the soil provides for the living organisms that live in the soil (micro and macrofauna) or depend on it (plants). Its major purpose is to produce the SIRS-Soils of the study area, gathering all the information concerning the definition of soil units of the area considered and the hydrostructural functioning of their representative pedon.

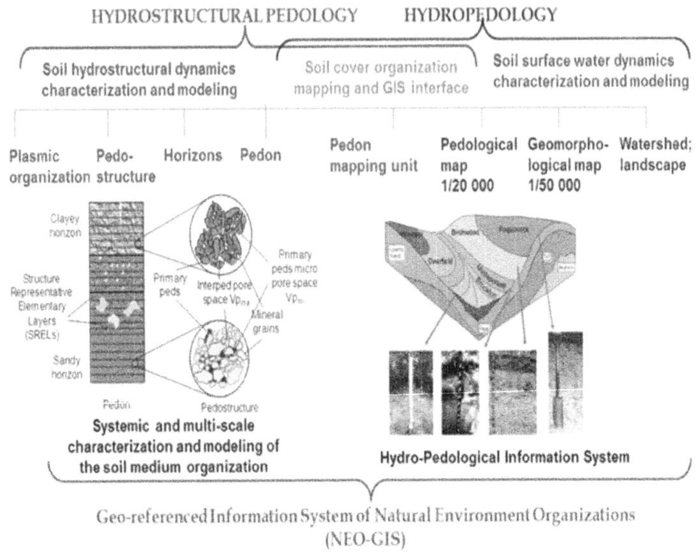

Figure 5.1. *The place of hydrostructural pedology in the Earth Sciences. The NEO-GIS is based on the SIRS-Soils used as the base layer, gathering more information about the environment (plant cover, microclimate, irrigation system, drainage). For a color version of this figure, please see www.iste.co.uk/braudeau/hydro.zip*

The association with hydropedology on the one hand and soil biology on the other, opens up two new research and development opportunities in the agricultural and environmental sciences:

– The methodological generalization of the of Natural Environment Organization Geo-referenced Information System (NEO-GIS, Figure 5.1) to all of Earth's soil-climate situations. As its name indicates, the NEO-GIS would contain all georeferenced information required for the systemic modeling of the "critical zone" (see Figure 2.2) of a geographical area, whether natural or not. Since the first piece of this information is the primary soil mapping unit provided with its representative pedon, the NEO-GIS is built and developed from the SIRS-Soils of the geographic area considered, which acts as the base layer so that other information about the "critical zone" at the soil surface, such as plant cover and microclimate (or local climate) have a well-defined relationship between the soil mapping units. Thus, due to its systemic construction, the NEO-GIS is the common ground of all couplings of agro-environmental disciplines with the science via the SIRS-Soils and the computer model Kamel satellite imaging with regard to the definition of units and information about the soil surface, land use, agriculture and ecology directly linked to the hydrostructural properties of the pedon and the corresponding soil unit.

– The study of the biochemical and biogeochemical properties of the soil in the laboratory and on the field *in situ*. All disciplines whose object of study is an organism (or group of organisms) living in the soil or dependent on it are concerned since, for the first time, the physical environment in which these organisms develop is defined as an *organized thermodynamic system*, whose micro and macro state variables are determined according to the water content [BRA 14a, BRA 14b], as shown in section 3.3. This allows the possibility of experimenting and controlling the "soil environment" in the laboratory, and therefore, monitoring and studying any biological process *in natural matrical conditions that the soil environment offers in terms of pedoclimatic regime pedostructural space and water.* This is highly important to soil biology laboratories which, without any available tool to characterize and model the hydrostructural properties of the soil, cannot generalize the results obtained in the laboratory on certain soils to the other soils or under other environmental conditions (hydrological regime, climate, geomorphological situation, etc.).

There are many research opportunities with respect to this second point: for example, using column of soil aggregates with known hydrostructural properties, determining the coupling parameters of the equations of the biological activity or chemical process analysed under controlled pedoclimatic conditions; or even the accumulation of pollutants in the inter or intra primary aggregate space, according to the hydrological regime that induces exchanges of water between the inter aggregate the microporosity and the macroporosity, all this followed and calculated by Kamel®.

So, to totally fulfill its role, the discipline must provide three laboratory units, specialized to process the objects of study unique to the discipline (in red in Figure 5.1); the studies will be automatically federated in the same paradigm (physical and systemic) represented by Kamel®:

1) a specific laboratory with standard equipment and methodology needed to measure the characteristic curves of the pedostructure (shrinkage curve, retention curve, hydric conductivity and swelling rate) and to obtain some other parameters at the pedon scale required by the Kamel® model.

2) a modeling unit developing the software Kamel® in different aspects of interdisciplinary coupling, both for the laboratory monitoring of coupling

experiments and for the simulation of agronomic production at field scale, using information from SIRS-Soils or NEO-GIS.

3) a unit for soil mapping, linked to satellital and geophysical survey, to first of all establish the SIRS-Soils that contains all information associated with the systemic definition of mapping units and the hydrostructural characterization of the representative pedon. As we know, this information allows the soil units to be modeled by Kamel®, which can be coupled with biological or agronomic models.

Units 2 and 3 (modeling and mapping) have no new features other than working in accordance with the systemic paradigm; their functions are defined in sections 3.3 and 3.4. The laboratory unit, however, must have a particular organization as we will present below.

5.2. Specificity of the hydrostructural pedology laboratory

The laboratory (Figure 5.2) is organized so that it meets its two research and development objectives mentioned above, that is to say: 1) meet the needs of systemic cartography to develop the SIRS-Soils, and therefore, characterize soil units for their hydrostructural modeling using Kamel®; and 2) meet the needs of experimental research on biological and biogeochemical soil processes under controlled pedoclimatic conditions, also at the pedostructure level in soil column as well as at the pedon level in lysimetric case.

The laboratory includes a part for the characterization of the pedostructure and part for the hydro-physical experimentation in soil column or lysimetric case.

5.2.1. *Hydrostructural characterization of the pedostructure*

The specificity of the laboratory lies in its equipment: a set of measurement apparatus specially designed to determine hydrostructural and hydrodynamic parameters of the pedostructure, characteristic soil parameters directly used by the Kamel® model. The main and vital piece of the apparatus is the TypoSoil® device [BEL 13], a new design tool that continuously measures, on eight samples at a time, the two characteristic soil moisture curves: the shrinkage curve $\bar{V}(W)$ and the water retention curve of the pedostructure $h(W)$. This device, the methodology and the associated processing data for determination of hydrostructural parameters of the pedostructure are described Braudeau *et al.* [BRA 13] and Assi *et al.* [ASS 14]. Part 2 of this book describes in detail the procedure used to determine these parameters in different existing cases (depending on the availability of measurement data), using Excel spreadsheets dedicated to each curve. The TypoSoil® device is pivotal as it is currently the only device that can provide all the information required to interpret results from other measurement devices. This is the case, for example, when determining the parameters of the hydric conductivity curve ($k_{ps}(W)$) as shown in section 8.3.5. or even the water absorption velocity constant by primary aggregates (k_{mi}) [BRA 06b].

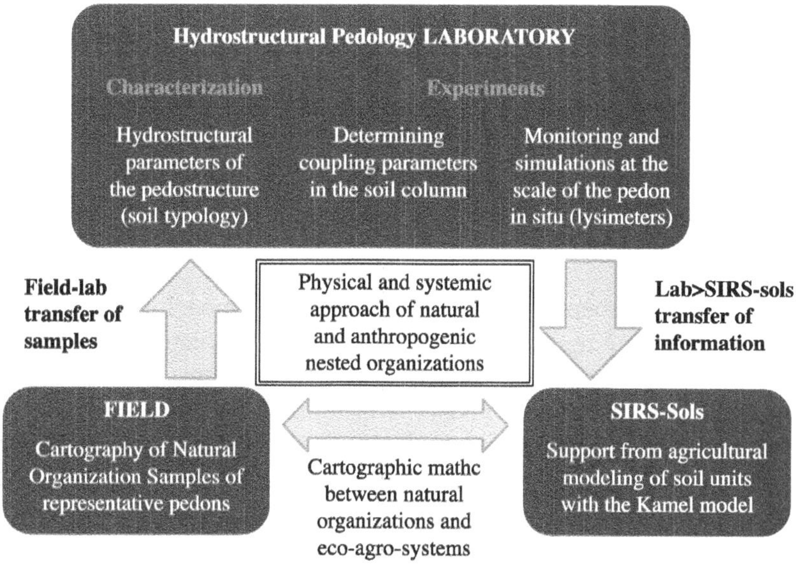

Figure 5.2. *The hydrostructural pedology laboratory as intermediary between recognizing soil organizations on land and the information system (SIRS-Soils) of the new discipline*

The main function of the laboratory with respect to hydrostructural characterization of the pedostructure is to provide to the SIRS-Soils all information about the soil units and their characteristic parameters of hydrostructural functioning, which will be processed by the Kamel® model. Since, with this characterization of the pedostructure, it is now possible to model the representative pedon of the unit according to the same internal organization found in the soil unit *in situ*, it is essential to retain the samples in the cycle of operations represented in Figure 5.2, from sampling in the soil horizons passing by characterization and experimental analysis in the laboratory, up to elaboration of the SIRS-Soils and its use.

With regard to other soil characteristics such as those of the pedostructure, which are required in order to model the soil-plant-atmosphere system with the Kamel® model, they are conventionally measured or estimated at the pedon scale in the field [BRA 14a].

5.2.2. *Experimental analysis of the bio-soil association in the soil column*

The laboratory "experiment" part (Figure 5.2) is also innovative as it can only exist if the soil, the physical environment of the processes studied being investigated, is previously characterized in its hydrostructural functioning. Using these characteristics, not only the pedoclimatic conditions imposed on the process can be controlled and programmed but also the state variables of the environment can be accurately determined at each instant and modeled with the Kamel® model.

We can devise adapted versions of the Kamel® model and its interfaces in order to simulate associations between the pedostructure and biological systems while monitoring their progress through the experiments. Comparing the experimental results with those from the simulation of the pedostructural and pedoclimatic soil dynamics using the Kamel® model coupled to the model of growth and production of the biological system studied, is a previously unseen way of obtaining parameters of growth and production functions of the biological model.

5.2.3. *Simulation of processes at the scale of the pedon using a lysimeter*

The same goes for experiments at the scale of the pedon using a lysimeter: previous knowledge about the pedostructural characteristics of pedon horizons allows the use of the Kamel® model to simulate the hydric functioning of the pedon. It gives the exact hydrostructural equilibrium at any point in the soil at any given moment, as well as flows of water in different horizon porosities [BRA 14a], and under field conditions (plant cover, PET, rain). It is, therefore, possible to consider what was not previously possible:

– the physical simulation of coupling experiments as above but at the scale of the pedon (succession of different horizons) and site conditions.

– interpreting the biological and chemical analyses performed in the laboratory of soil samples taken when monitoring the quality of water and soil in hydro-agricultural systems. The systemic modeling of soil using the Kamel® model is the only modeling method in place to reproduce laboratory data under soil conditions *in situ*.

6

Implications for Agro-environmental Sciences

6.1. A unitary theory on the systemic and thermodynamic approaches within the natural environment

6.1.1. *Modeling in the natural environment and Cartesian precepts*

Speaking about problems inherent to soil science (section 2.3), we had noticed the absence of a unitary theory for resolving the two fundamental problems raised, which are intrinsic to soil science: the recognition of the levels of hydrofunctional organization in the natural environment and the physical formulation of hydrofunctionality of the soil's organized and structured medium.

These two problems have been resolved by inscribing soil science into *the new descriptive paradigm of systemic soil modeling*: hydrostructural pedology. It results in the complete fundamental theorization of *the systemic approach of the natural environment* that we have used here, starting from the works of Le Moigne [LEM 94] on the definition and the modeling description of what "should be" the General System.

Adhering to this idea of a General System as a model of a universal system on the relationship between man and nature, as foreseen by Bertalanffy, we have taken certain materials described by Le Moigne, but re-inscribed these into the Cartesian logic that he "radically" abandoned in 1976 (first edition of his book), when he posed "The four precepts of the new discourse of the method" [LEM 94, p. 42] in place of the with the four precepts given by Descartes (1637) [LEM 94, p. 30]. One of these in particular, which Le Moigne replaced with that of aggregating, is exactly that which could lead to the notion of an Organized Natural System, as we have defined it, based on the SREV notion (section 3.2.1), which can be written as: $V_{tot}=\sum V_i$, where each V_i is determined. This precept was enunciated like: "And the last one (precept), to make everywhere such complete enumerations and such general reviews that I can be sure not to have omitted anything".

The aggregating precept as defined by Le Moigne, allows on the contrary the modeler to go beyond this last precept, which can only lead to the REV notion, or to the black box where the internal structure (organization of the solid phase) of the object is overlooked, as Le Moigne [LEM 94, pp. 40–41] clearly says:

> "*The last precept of the former discourse, on exhaustiveness, can readily be rejected. It is so shamelessly overridden on a daily basis by everyone, Cartesian or not, that its defenders will not resist too much when faced with the factual argument: it is... in practice... impracticable! Who could ever be sure that they have made such a complete enumeration, that they are sure to have omitted nothing*".

And therefore, further on (p. 41):

> "... we can no longer agree that we too, *shall "make" such complete enumerations and such general reviews, that we are sure not to have omitted anything*".
>
> *It is better to agree and to deliberately propose that we have omitted many things by pushing them into the shadow of aggregate; aggregates that, of course, we will select ourselves, both explicitly and openaly. From this point on, we will no longer claim to explain 'everything' about the object of study (with some risks...)*".

We have seen that it is this way that the hydro-physicians working in soil science took for the modeling of water transfers in the natural environment, when they came up with the REV hypothesis to free themselves from the constraints linked to multi-scale soil organization and the apparent heterogeneity of the structure. This organization was deemed too complex to be described in a mechanistic way, and from there arose their adhesion to the black box concept, developed by cybernetic modeling. In so doing, the researchers resigned from the traditional idea "of a natural perfect order", abandoning not only the fourth, but also the second and third of Descartes' precepts, which are: (2nd) "to divide up each of the difficulties which I examined into as many parcels as possible, and as seemed requisite in order that it might be resolved in the best manner possible". (3rd) "to carry on my reflections in due order, commencing with objects that were the most simple and easy to understand, in order to rise little by little, or by degrees, to knowledge of the most composed, even, assuming an order between those which do not follow a natural sequence relatively to one another".

These precepts were also rejected by Le Moigne. He qualified them as reductionist precept and causal (deterministic) precept, question them vehemently by bringing out their dogmatic and restrictive side for the modeler:

> "We will no doubt suffer a long time still from this imperialism implicit in the third precept, and it will be difficult to convince us that is possible to be perfectly rational without being constrained to the only causalist model for knowing the world".

The two precepts that he proposes in replacement in his new discourse on the method are:

– *The globalist precept:* henceforth, perceiving the object to be known as a part that is inserted, immersed and active in a larger whole (we would soon say: in an environment), and to make the intelligence of this environment the prerequisite to our knowledge of the object. That is the extent of the new precept to which we will oppose reductionism: it will be recognized under the label of globalism.

– *The teleological precept:* to interpret the object, not in and of itself, but by its behavior, without looking to explain this behavior a priori, by some implied law in a potential structure.

We will in fact see that the systemic paradigm we devised here, not only is the adequate methodological response to the requirements of the Cartesian precepts, which Le Moigne found impossible to respect, but also resolves all issues raised in critique of the later. So if there is a new discourse on the method to be issued, it could be that of Descartes on condition of adding this precept: working in the systemic paradigm.

Indeed, the results that we have obtained from the systemization of a natural organization (soil) allows us to put forward the following remarks:

> The second precepts of both Descartes and Le Moigne (reductionist/globalist) are not at all contradictory, to the contrary: they complement each other. Descartes' analytical method is obviously necessary and should not be discarded; however, he missed the two systemic concepts that we introduced here: 1) the system's definition as a delineated space with a closed structure, but open to liquid and gas fluxes; and 2) the systemic description repository, being the three scaled axes that allow the precise recognition and description of the "parcels of problems" mentioned by Descartes and which today we would call functional organized systems.

With the contribution of these two concepts, Le Moigne's critique that it is a reductionist precept that does not consider the external global aspect no longer stands. In effect, with the system defined as having a concrete external delineation that incorporates a solid phase, the global system's external delineation constitutes the functional interface with its external environment. On the other hand, thanks to the idea that the subsystems' (on one same level) complementary volume within the global system (assembly of these subsystems) is exactly that of the liquid and gas phases surrounding the solid phase and not retained by the systemic delineations, all of the internal functional volumes in the global system are necessarily enumerated and determined. For example, if S1 is the only subsystem of S3, scattered in it but of total volume known (V1), the complementary volume (Vp) is not a system as it is only occupied by the mobile phases in S3, it is exactly determined by the volume difference of the two systems (V3−V1). Therefore, the systemic analysis gives an easy solution to the "impracticability" of Descartes' fourth precept (enumerating everything) according to Le Moigne. Finally, the Le Moigne's aggregation precept should be completed in the systemic paradigm in order to join that of Descartes and avoid the use of the black box (or more precisely REV) principle.

With regard to Descartes's third precept (*to carry on my reflections in due order...*), on the condition of using the systemic description repository concept, defined above (3.2.2) for the systemization of natural organizations, this precept corresponds exactly to the approach that we followed for building the computer model Kamel®. Based on the concepts of pedostructure and SREV,

Kamel® models the hydrostructural functioning of a pedon as a representative of a primary mapping unit of soil. This approach in no way refers to this teleological precept proposed by Le Moigne to replace this third precept.

In fact, we can say that this teleological precept is in contradiction to Descartes' first precept: "to accept nothing as true which I did not clearly recognize to be so: that is to say, carefully to avoid haste and prevention in judgments, and to accept in them nothing more than what was presented to my mind so clearly and distinctly that I could have no occasion to doubt it". The teleology precept should allow the modeler to dispense with Descartes' 3rd precept and so to base his modeling on the imagined role of the object at the heart of its environment, rather than on the relationship of cause and effect (to be researched) between interior and exterior of the object, implying the internal organization of the object. Resorting to this precept is understandable when the object is manufactured by man; in this case we can effectively hope to understand its internal mechanism based on its role or utility in the social external environment. However, when the object is natural, the modeler observing this precept will necessarily pose teleological hypotheses which are impossible to generalize and deepen.

As for us, we have responded to the first of Descartes' precepts by making a fundamental distinction from the beginning, between organization and system: an organization is a divine creation, whereas a system is a human creation. So we have to transform the natural organization into system for being allowed to study it. We have given the rules of transformation into systems of all natural organizations and systems, what we call: systematization. Thus, we can understand why there can be no room for teleology in our "natural system" as physicaly defined here, and, in the end, why the Descartes' four precepts are the basis for the *physical* systemic approach.

6.1.2. *Systematization, theoretical basis of the systemic approach*

The operation of transforming an organization into a system is called "systemization" of the organization (study object) to be described. As shown, this systemization is done in two stages (Figure 6.1). The first stage is simply the concrete delineation of the object to study: a soil profile, a geomorphological area, a region or other naturally organized body. This delineation precisely demarcates the interior from the exterior of the object and entirely envelops the whole solid phase, which is blocked in the internal structure. The object has become a system that we can bring to the study as an OS for the GS (the scientific discipline) of which it is, on its own, the object of study. The second stage of systemization (Figure 6.1) must be performed on the object in order to transform it into an organized system. This stage is absolutely necessary as it *exhaustively defines* the descriptive variables of the OS's internal organization, at all recognized hydrofunctional levels. These will be the exhaustive variables used in the physical modeling of the real object's internal and external activity. This is done by describing the system object according to the SREV concept in *the systemic descriptive repository of the three graduated axes,* presented in section 3.2.2. The first stage corresponds to the REV principle and the second to the SREV principle.

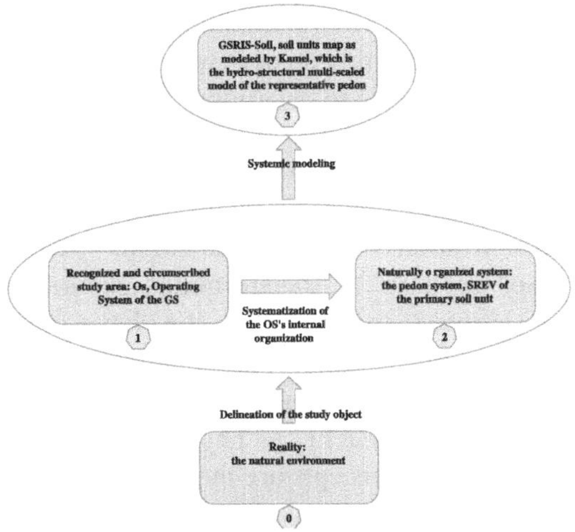

Figure 6.1. *The different stages of creating a SIRS-Soils (Soil Information System) on hydrostructural pedology*

Yet it is this concealment of structure and the natural object's organized hierarchy by the use of REV, in which the complexity of the said "complex systems" and the empiricism of mathematical equations are born, to which we attribute the REV (or black box) functioning and activity, regarding its external environment. The complexity that "complex systems mathematicians" speak of is, in fact, one of organization, the study object's natural internal organization, not of the corresponding organized system: the corresponding SREV defined by man to study this organization All systems created by man may be complicated, but *in essence, cannot be complex.*

Thus, the complete systemization of the natural object resolves the problem of the apparent complexity of its internal organization by transforming the same into a system, organized into subsystems, which are themselves organized into subsystems. This is done hierarchically until the final subsystems become objects of the solid phase, which constitute the plural-scaled organization structure. The soil's living organisms: plants; roots, macro-fauna; etc. are considered autonomous systems that share space, air, water, and heat with the pedostructure in the internal soil organization, represented by its pedon.

6.1.3. *Theoretical and applied systemology*

We want to show here that the systemization operation of the natural object illustrated in the Figure 6.2, follows the same pattern as systematicians such as Mesarovic, Klir and Orchards, as analyzed by Pouvreau [POU 13]. They made important contributions to developing the "theoretical fundamental systemology", initiated by Bertalanffy, and worked particularly on the logical and methodological codification of the "general systems" theorization. Regarding the procedure of "definition of a system on an object", Pouvreau [POU 13, pp. 940–942] notably explains that:

> In this description, one must take into account the fact that Klir distinguished the "study object" from the "system defined on this object" (that is to say, from a systemic model of this object), much more clearly than Mesarović. The essential principle of the procedure that he described is that this system, as much as a "source system" (defined as a group of system variables and a group of states associated with these variables) is created by both: homomorphism to an "object system" and isomorphism to a "general system". (The object system is a construction of the object that I qualify pre-systemic, defined by the group formed by selected object attributes and of appearance sets of each of these attributes). This point underlined by Le Moigne in his 1977 "systemography" theorization, was already done so by Klir two years beforehand.

> But it was Orchard who, from 1972 onwards, would contribute most significantly to the systemization of what Klir called the "procedure of defining a system on an object", although he endeavoured to synthesize and complete Klir's works. Orchard insisted on the stages that allowed allowed one to connect a specific problem to a "general system", for the purpose of constructing a systemic model destined to resolve the problem.

After having recalled the methodological propositions of these authors (1970–1980), Pouvreau [POU 13] presents the updated synthesis by the diagram in Figure 6.2: It represents the construction stages of the study "systemic model of the object" of study, passing through that of "object system" (numbers 1 and 2) and of "systemic object-model" (number 3) under the general systems umbrella, and then to the modeling per se, which links the "real sciences" to the "general systems theory":

> It seems to me not only possible, but useful and even necessary to further refine Klir and Orchard's descriptions of the systemic model construction procedure. The distinction I made in the second part between "object model", modeling and model, of which I illustrated the pertinence, while I considered Bertalanffy's growth theory, becomes particularly useful here.
>
> As it is not directly the "system-source" (or source system) which Klir talked about – that the systematician models by isomorphic construction to the "general system"- but an abstract representation, linking the selected variables in a very formal and unspecified way: this representation is what I call the "*object-systemic model*" of the study object. Klir's idea of distinguishing between a "*system-object*" and a "system-source", and to consider their relationship as an homomorphism from the first to the second, seems to me more precise in contrast to Orchard's very vague approach to the first one. Orchard's "system-object" seems to blend three distinct concepts (the two distinguished by Klir and that of "*object-systemic mode*") into one. It further seems necessary to specify that the "*general system*"'s direct function concerns the construction of the object to be modeled, whereas that of the "*general system*" theory, concerns the actual modeling. Therein lies its function as a logical guide.

In this diagram, we can recognize the exact correspondence between the numbered objects of Figure 6.2 and those of Figure 6.1 earlier. They represent the elaboration process of SIRS-Soils by hydrostructural pedology, a discipline organized by isomorphy of the general systems model, from the scientific disciplines in the natural environment that we have defined here.

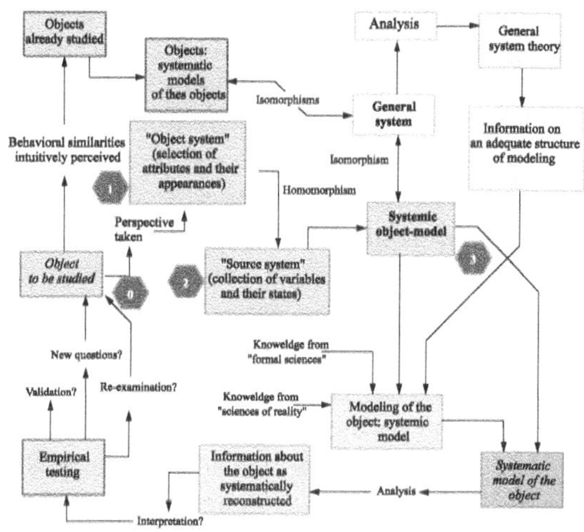

Figure 6.2. *Reconstruction of the diagram showing the principle functioning of "applied theoretical systemology", according to Pouvreau [POU 13, p. 943]. For a color version of this figure, please see www.iste.co.uk/braudeau/hydro.zip*

Starting from the "*object to be studied*" [0], the "system object" [1] corresponds to the real object, which was delineated (OS of the GS) and of which we empirically know certain properties: "*system-source*" [2] is for us the representative pedon from the mapped soil unit, a system organized into subsystems, themselves organized into subsystems, etc. The descriptive variables of these are all itemized The "*systemic model object*" [3] is the soil map proceeding from a systemic cartography of the landscape, which supplies information on 1) map unit demarcations and 2) the characteristic parameters of soil units, namely the physical modeling parameters of the representative pedon's hydrostructural functioning, as contributed by the Kamel® model.

Then, after [3] comes the "modeling *of the object-system model"*, which comes from the Kamel® model [BRA 14a] and is a simulation of the "study object's" internal and external activity (and potentially also of its correlated structural evolution). The representative pedon is submitted to the activity of its internal environment (plant organisms and biological micro-organisms) and its external environment (climate, cultural materials, culture, etc.) This simulation of the "soil" object in its climatic biological and agronomical environment requires information on the environmental state, contained in what we call the NEO-GIS. This is a (systemically) constructed geo-referenced information system based on the SIRS-Soil of the area in question (see Figure 5.1). The *"systemic model of the object"* is the result of the simulation over a given period of time on the production, water consumption, evolution of the layer, etc., on a hydro-agricultural area, for example. As mentioned in Figure 6.2, the results of the simulation in the given time are then analyzed, compared and contrasted to reality to be interpreted, and to respond to the questions, etc. We note here that if the notion of the physical system is respected at each stage, that is to say, if we have not brought back empirical (not systemic) variables in the modeling process, there is no place to research the validation of the model itself. The Kamel® model, because it is systemic, exactly reflects the reality of the soil and is, therefore, generic: if all the required information is given, it functions in the same way on all soil types and in all environmental conditions without need to be calibrated.

In the upper part of Figure 6.2, one notes that the "*general system*" is in an isomorphic relationship with the "*object-systemic model*"; it is on this point that we distance ourselves from the general system definition or vision of the authors analyzed by Pouvreau, while remaining consistent with the perception of this last given in the diagram. The natural system (NS) representing the "general" organization of the soil-plant-atmosphere continuum, which we have physically defined as the representative pedon of the map unit of soil and that Kamel® is potentially capable of modeling in all internal and external aspects of behavior and evolution, at all hydrofunctional levels, could be called the "General System of the natural environment" [POU 13, p. 917]. However, in imposing from the beginning, the "systemization" of the study object as a prerequisite to every systemic construction model of the object, with the notions of 'system" and "organized system" that we have *physically* defined here, we eliminate from the beginning, the possibility of a plurality of non-systemic models for representing the same phenomenon. These are models that rest on the principle of the black box, ignoring the internal organization and, therefore, unitary physics. In fact, any systemic model, in the physical sense given here, can be said unitary for it is described in a unitary way in the repository of the three systemic axes that have been defined here; and will remain such, so long as we have not introduced empirical notions in its hydro-physical internal functioning. If there is only one Natural System (NS), representing the general organization of the natural environment, there will, therefore, only be one systemic model to represent it. It is this NS that could effectively be qualified as general system of nature; however, because each system is a human creation, it is better to reserve the appellation of the General System to the scientific discipline (Figure 4.3) that is built according to the general system theory and in charge of the natural environment study. With its three component subsystems (SP, IS, OS), it systemizes the internal organization of the object to be studied in order to have the knowledge , the representation and the total control of it. Thus, we continue to conform to Pouvreau's [POU 13, p. 142] diagram (Figure 6.2) which tells us "the direct function of the general system concerns the construction of the object to be modeled while that of the "general system theory" concerns the modeling, strictly speaking". The discipline in question, in the hydrostructural pedology occurrence, uses its own systemic description repository (three graduated axes) to guide the systemization of the natural object to describe, define its descriptive organization and functioning variables and to model it, as explained in sections 3.2 and 3.3.

As a consequence, the "*general system*" placed above the "*object-system model*" [3] (Figure 6.2) is the GS of hydrostructural pedology, as defined in section 4.4 and shown in Figure 4.3. Its information system is *in a relationship of isomorphism* with the "object-system model", which is, in our case, the soil mapping units, supplied with its representative pedons that can be modeled by Kamel®. This information system is what we call the SRIS-Soils, specific to the GS of hydrostructural pedology.

In order to be complete in the comparison of Pouvreau's diagram (Figure 6.2) with the methodology of SIRS-Soils elaboration, we draw attention to the fact that the "*systemic model objects*" of the objects already studied (in the top left of the figure), which are in an isomorphic relationship with the SG, are SIRS-Soils established in the same systemic methodology in diverse regions of the earth. They would then be juxtaposed with full compatibility of scales, borders and representation models.

This SIRS-Soils extension to all the terrestrial surfaces would make of it the virtual image of the physical soil environment, defined and presented in each point in space in terms of thermodynamic equilibrium and hydrostructural states; and within which the soil's physic-chemical and biological processes that are caused by living (active) organisms, could be modeled. The coupling modeling of biological life with the soil must be done on the *'modeling of the systemic object model"* case represented in the diagram. That is to say that they must be foreseen and addressed by the Kamel® model.

6.1.4. *General Systems theory of the agro-environmental disciplines*

It is noteworthy that the principle diagram of applied theoretical systemology raised by Pouvreau (Figure 6.2) is comprehensive and in complete agrees with what we have effectively carried out and theorized terms of cartography, characterization and agronomic modeling of the soil in the systemic approach, for the last 15 years [BRA 01, BRA 07, BEL 08, BRA 09b, SAL 12].

Our contribution to the general systems theory is mainly the three concepts put in place to carry out the systemization of the study object: 1) the physical definition of natural system (GS's operating system) that differentiates it from the natural organization of which it is a representative extract; the natural system is a thermodynamic system, which is closed off on its structure, in agreement with the SREV notion. 2) The concept of "systemic description repository" with its three graduated axes, to transform the natural system's internal organization into organized subsystems and 3) The concept of the General System as the human system of a scientific discipline organized as the isomorphic image of the evolved system of level n°9 from Le Moigne's [LEM 94] and of which the study object is the natural environment. We recapitulate these three points below:

1) The natural system: The systemization of the natural object and its internal organization led us to the definition of an organized system that has the property of being a thermodynamic system closed off to the structure (solid phase) of the organization it represents, but that stays open to mobile materials that flow into this structure. Its descriptive variables stem from the SREV concept and then can be said to be systemic. The exchanges between the systems of space, heat, water and nutritive elements, are thus governed by thermodynamic laws.

2) The systemic description repository: We have defined it as being the intellectual approach that allows one to describe an (organized) natural object as a natural system, concretely delineated and positioned in the georeferenced space, according to a triaxial description repository in which the three scalable axes, respectively, represent:

Axis I, the structure of organizations and their morphology assembly at different hydrofunctional ranges (spatial occupation of organizations, range levels)

Axis II, the evolution of defined organizations on axis I, their genesis and their end (stages of evolution)

Axis III, the internal and external activity of defined organizations on axis I, to a given stage of evolution on axis II (simulations functioning in common time).

The description on the three axes fits the two types of natural objects that we must distinguish: the living organism and the physical environment in which this organism develops. Each of them must be described as an organized system in this same description space to the three axes. This ensures their relational compatibility in terms of occupation of and thermodynamic equilibrium of activities.

3) The General System model of a scientific discipline: We have shown how man's relationship with his study object was perfectly represented by Le Moigne's (Figure 3.2) GS diagram, adapted to the agro-environmental sciences (Figures 4.1 and 4.3), so that all scientific disciplines in the natural environment can be considered a system constructed by man in an

isomorphy of GS. The triaxial datum of systemic description that we have established to describe the OS of the GS is the discipline's methodological tool, which allows it to erect an information system (IS) in conforming with the systemic paradigm. The objective of the discipline, once it has been defined as isomorphic to the GS, is to grow in science, that is to say in knowledge of the object (OS), and to formalize this knowledge in a way so as to register it in the IS, the OS's systemic and virtual image. Seeing the diversity of the situations and observation ranges, this growth in science can only increase while remaining in the same paradigm. This is the systemic paradigm, which is, as we have seen with the unitary paradigm, not limiting, where the living organisms and the environmental sphere in which they live are represented by systems in relationship of thermodynamic and hydrostructural equilibrium.

Therefore, the General Systems theory is for us a theory on man's relationship to nature, study object, exploitation or even management, which links it to Bertalanffy's ambitious project to elaborate a general systems theory. Pouvreau [POU 13] understands and describes it as a general science of systemic interpretation of the 'real". The recovery of Le Moigne's [LEM 94] concepts of General System and systemic triangulation, but which we have modified in order to place them back into the Cartesian paradigm of not blending that which belongs to the divine (natural organization) and to the human (organized system) has brought us to the notion of a thermodynamic system in a natural environment that we will present below. It represents a qualitative leap in the paradigm of the systemic approach by unifying the systemic approach to the thermodynamics of the natural environment.

6.1.5. *Systemic thermodynamics*

At every range of the agro-environmental sphere, natural systems' functionality is envisaged in relation to water. The selected organization levels during the systematization of the object system 1 to the source system 2 are always hydrofunctional (axis III), which makes the relationship from 1 to 2 always homomorphic (in respect of the boxed ranges). The tri-phased soil–water–air equilibrium, on the organization level of the pedostructure, is the hydrostructural equilibrium, according to water. They are represented in their totality by the pedostructure's shrinkage and retention curves in the soil sphere. These two curves form the two hydrofunctional characteristics of soil structure.

The systemic definition of pedostructure [BRA 04], that is to say, is the definition of the pedostructure as an organized natural system and, then, a thermodynamic one, has represented a major advancement in the physics of soil water. It has allowed the exact thermodynamic formulation of equations on the shrinkage and retention curves of soil water, which were reputably unknowable due to the complexity of the soil sphere. These equations rest on a fundamental hypothesis in the thermodynamics of soil water, which has been totally validated by experience [BRA 14c, BRA 14d]: to know that Gibbs' free energies of aqueous phases of the pedostructure: $\bar{G}_{wmi} = W_{mi}\mu_{mi}$ and $\bar{G}_{wma} = W_{ma}\mu_{ma}$ (in J/soil kg) of equation [3.2] *are constant with the variation of water amount.* That brings about a very simple relationship between the micro and macro amounts of water with their respective potential:

$$\mu_{wmi} = \bar{G}_{wmi}/W_{mi} \text{ and } \mu_{wma} = \bar{G}_{wma}/W_{ma} \qquad [6.1]$$

which, associated with the highest equation [3], has given the exact equation of the water retention curve: $h(W) = h_{mi}(W_{mi}) = h_{ma}(W_{mi})$ (see section 3.3.2).

It can be stated that the relationship $\mu_w = \bar{G}_w/W$ is not valid as the potentials cannot be averaged: $\mu_w = \mu_{wma}$ in the macro-aqueous phase and $\mu_w = \mu_{wmi}$ in the micro-aqueous phase. The exact equation of $h(W)$ was to *never be found for as long as one ignored the existence of the two types of internal and external water to the primary aggregates,* W_{mi} and W_{ma} that is to say for as long as one did not take into account the hierarchical organization of soil sphere. These two types of water must necessarily appear in the total and exact differential equation of Gibbs' free energy, instead of the habitual term (μdW):

$$d\bar{G}_w = -\bar{S}_w dT + \bar{V}_w dP + \mu_{mi} dW_{mi} + \mu_{ma} dW_{ma} \qquad [6.2]$$

The fact that we find $\bar{G}_{wmi}$ and $\bar{G}_{wma}$ to be constant despite the changes to the amounts of water is due to the fact that the distribution of charges on the clay surface in the primary peds (micro) and to their surface (macro) stays the same, independently of the amount of water [BRA 14c]. These charges that develop a field of potential in the liquid phase, coating the solid phase, are directly responsible for the soil's water retention *h*. This retention pressure corresponds exactly to what we in soil science call the "matrix potential, ψ_m" without having been able to give a physical explanation.

The constancy of free energies in the pedostructure's micro and macro ($\bar{G}_{wmi}$ and $\bar{G}_{wma}$) phases with the changing amount of water brings about a particular writing on fundamental equations of thermodynamics relative to the pedostructural water:

First of all, $\bar{G}_w$ being equal to the sum of ($\bar{G}_{wmi}$ and $\bar{G}_{wma}$) is constant, and the total differential of Gibbs' free energy of pedostructural water will always be equal to 0, *regardless of the water amount*:

$$d\bar{G}_w = \bar{S}_w dT + \bar{V}_w dP + \mu_{mi} dW_{mi} + \mu_{ma} dW_{ma} = 0 \qquad [6.3]$$

On the other hand, the derivative $d\bar{G}_w$ under its Eulerian form, gives:

$$d\bar{G}_w = d\bar{G}_{wmi} + d\bar{G}_{wma} = d\left(\mu_{mi} W_{mi}\right) + d\left(\mu_{ma} W_{ma}\right) = 0 \qquad [6.4]$$

$$\mu_{mi} dW_{mi} + \mu_{ma} dW_{ma} = -\left(W_{mi} d\mu_{mi} + W_{ma} d\mu_{ma}\right) \qquad [6.5]$$

which leads to the Gibbs Duhem equation for the aqueous phase of the pedostructure, when rewriting equation [6.3] according to [6.5]:

$$\bar{S}_w dT - \bar{V}_w dP + W_{mi} d\mu_{mi} + W_{ma} d\mu_{ma} = 0 \qquad [6.6]$$

Finally, knowing that the internal energy of the pedostructural water is written as:

$$\bar{U}_w = T\bar{S}_w - P\bar{V}_w + \bar{G}_w \qquad [6.7]$$

its derived form is:

$$d\bar{U}_w = Td\bar{S}_w + \bar{S}_w dT - Pd\bar{V}_w - \bar{V}_w dP \qquad [6.8]$$

and by adding equations [6.3] and [6.8] term by term, we obtain the classic formulation of the differential form of internal energy:

$$d\bar{U}_w = Td\bar{S}_w - Pd\bar{V}_w + \mu_{mi} dW_{mi} + \mu_{ma} dW_{ma} \qquad [6.9]$$

with the difference that the habitual term $\mu_w dW$ is necessarily replaced by the two relative terms of the two micro and macro a phases of the pedostructure, that is to say, internal and external to the primary aggregates.

In conclusion to this section, we can say that it is the first time that an exact terminology and systemic thermodynamic formulation has been used in the pedoclimatic atmosphere of one point in soil. We have exhaustively used the Kamel® model as an intermediary for all the variables of the soil sphere state, locally at the range level of processes, which enter into coupling relationships with biological organisms. They are in a relationship of exchange with the soil, in terms of water, air and heat fluxes.

We recall that thermodynamics are said to be systemic when their batch of descriptive variables is exhaustive and defined by its reference to the object's internal organization Variables such as density and partial pressure are not systemic. It is easy to show that Gibbs' paradox dos not exist if instead of speaking about partial pressure, we speak about partial molar pressure V/N, the volume occupied by a mole in a group of N moles, occupying a well-defined volume V (in terms of temperature and pressure). Analogous to the "system closed off to its structure" (SREV) for the liquid and gas phases is the variable volume V enveloping a fixed number of molecules N. V/N is a systemic variable, like all other *extensive* variables applied to N, while N/V or P/N are not. In the gas phase for example, if Vi and Ni are the volumes and numbered of gas i molecules, for gas we have the perfect simple relationship: $V_i/N_i = \sum_i V_i/\sum_i N_i = V/N$ to which it has a physical meaning, which contrasts to P_i/N_i. Braudeau *et al.* [BRA 14c, BRA 14d] showed that it is by exclusive use of the systemic variables that we could establish the thermodynamic equations of the soil's water retention, and to show the equivalence of the two measuring methods, based on very different physical principles. The systemic definition of extensive variables in thermodynamics, the presence of content variables in micro and macro-water in the pedostructure's thermodynamic equilibrium equations (where all soils are concerned), the generality of equations [6.3]–[6.9], all of this can be viewed as a new contribution to the systems theory, or rather "to the *general systemology* project as it was formulated and developed by Bertalanffy, Rapoport, Boulding and those systematicians who inspired it –

Bertalanffy having first spoken in 1965 of general humanist systemology" [POU 13, p. 911].

In effect, by this thermodynamic system, we respond to the objectives of the open systems theory, expressed in the 1930s by Bertalanffy, namely (cited by Pouvreau [POU 13, p. 643]) "the need to conceive the organism as an open system and to find formulations which allow one to characterize these 'equilibrium fluxes', in a manner analogous to that in which known physical chemistry expressions define the genuine chemical equilibrium in closed systems".

We describe a natural system as an SREV organization in which a "piece", is completely compatible with Bertalanffy's idea of considering "the organism as a physical system" that is open to equilibrium flux with its environment. What no one saw at the time was the distinction to be made between an active, living organism and the natural physical environment in which it lives. It is this environment that is more visible than the organism, "ordered by successive levels of organization, to conceive in a hierarchical order of systems open to equilibrium fluxes".

6.2. The new challenge to agro-environmental modeling

The other important implication in agro-environmental sciences is the vital emergence of hydrostructural pedology. The challenge of today, which is made possible by hydrostructural pedology, is model living organism activity, each in the place it occupies in the hierarchized organization of the physical natural environment; the forming group is what we call the "critical area" (Figure 2.2). The physical system of organizations, which describe the material exchanges and the energetic equilibrium that govern them, between the organisms and their sphere of activity, becomes the interdisciplinary language of this challenge. It is in this same physical and systemic paradigm that we can experiment with and model the coupling of biological systems with the soil system, submitted to external climatic variations.

In Figure 5.1, we have seen how this new discipline of soil science, with its physical system, adjoins hydro-pedology, which is not systemic; and how also with its analysis tools and specific concepts (hydrostructural characterization, Kamel® model and SIRS-Soils), this new discipline comes to occupy a practically vacant place today in agro-environmental sciences. As hydro-pedology is incapable of modeling soil structure and its interaction with water, it cannot invest the same scale of work as its colleagues do, which are found in the functional levels of processes in the soil. In return, without the necessary localized information on the representative pedon and map units, hydrostructural pedology cannot invest the great environmental units used today (SWAT model, SHETRAN hydro-logical model, etc.) into the realm of macroscopic modeling (global scales, off-scale processes).

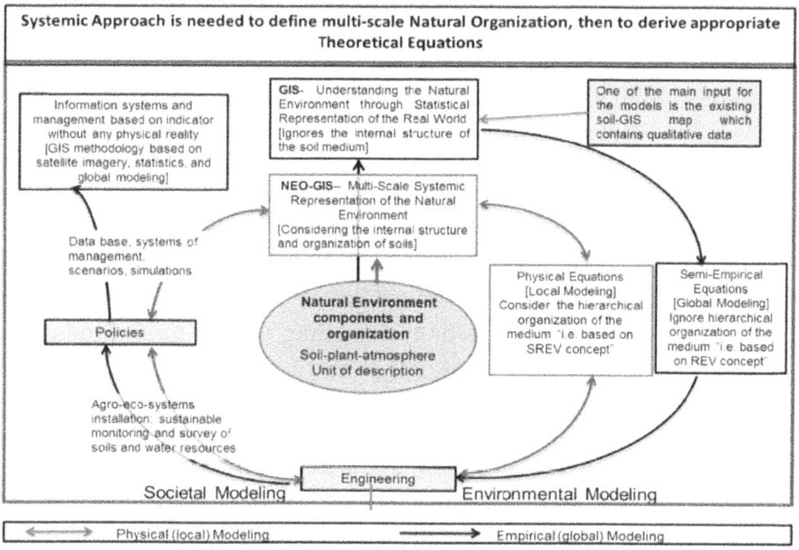

Figure 6.3. *The double cycle, local and global, of modeling the soil-water couple in the natural environment (ecosystems) and societal (agrosystems) (from Braudeau and Mohtar [BRA 14b]). For a color version of this figure, please see www.iste.co.uk/braudeau/hydro.zip*

Yet, these two modeling types are as necessary as each other in the support software when it comes to deciding or managing natural resources for agricultural development, which is as well adapted to the local scale as it is to the national or international scale. Braudeau and Mohtar [BRA 14b] have shown that it is possible to establish correspondence and a methodological line between the two modeling types practiced in the two disciplines by linking the macroscopic or global (mono-scale, statistic) description from one study area to the systemic, multi-scale and mechanistic description of the internal physical sphere of this area.

This could well constitute the future challenge of agro-environmental modeling. In fact, the modeling cycle used today for a country or region's agricultural development is only the global or macroscopic modeling cycle figured in black on the diagram in Figure 6.3. This cycle is entirely based on a non-systemic vision of the natural setting [BRA 14b]. Contrastingly, the cycle outlined in red in the diagram, which is the physical and systemic modeling one of the natural setting that we have developed here, could not be implemented as it required changing the paradigm to one which was not yet recognized

The first stage of the new challenge consists of installing the physical and systemic modeling of water and soil in the natural setting, in red in Figure 6.3; the cycle is then implemented and practiced by hydrostructural pedology. Contrary to the first one, this cycle is represented by double-sided arrows to show that it is on the local scale of the processes and mechanistic that is to say linked to the causes. All of the stages are explanatory and it is physically possible to establish an exact comparison of hydrofunctional properties in the environmental setting before and after the transformation it undergoes due to human action (installing agricultural production or environment management systems, left part of Figure 6.3).

This challenge is, therefore, heavily linked to the development of hydrostructural pedology, of which the scientific objectives coincide with the systemic and physical modeling of the organization of both the natural and social settings, which are to be implemented. To track environmental resources and implement agricultural policies, this physical cycle (or local cycle, in red) should be placed in a relationship of correspondence with the statistic model (or global model, in black) that is practiced by hydro-pedologists (right hand side of the graph) and by agro-pedologists (left side). The result (Figure 6.3) is the development of trans-disciplinary information in the NEO-GIS (Natural Environment Organization-GIS) platform, to systemically model what we call the "critical zone". For us pedologists, this is the mapped unit of soil, enriched by its vegetation cover and other organisms colonizing the soil, of which the hydrostructural functioning is completely represented by the representative pedon, such as the systemically modeled by Kamel®.

PART 2

Hydrostructural Characterization of Soil Pedostructure

This second part presents the measured data processing methodology for obtaining the characteristic parameters of the pedostructure hydrostructural behavior, defined and described in the first part (see section 3.3). In order to do this, we have detailed and clarified the programmed Excel spreadsheets used for determining and calculating the parameters of characteristic curves by adjustment of the theoretical curves on the curves measured with the use of the TypoSoil® apparatus that (see section 3.3.3) measures, simultaneously, both soil moisture characteristic curves: shrinkage and water retention curves, on a soil sample. However, because this apparatus is a recent development, and various curve measurements, which were obtained in different ways already exist in databases, we have also processed different cases where the characteristic curves were obtained without the use of TypoSoil®: the shrinkage curve measured alone, the water retention curve measured alone (using the tensiometer), the traditional pFs curve, transformed into tensiometric water retention curve and finally the case of the pedostructure hydric conductivity. In this last case, a new procedure for determining the parameters of the hydric conductivity curve was tested, using the new thermodynamic equations for the water retention curve. The Excel sheets used for determining all the pedostructure hydrostructural parameters and described in Chapter 9 will be made available to the reader by downloading in order to facilitate the understanding.

7

Introduction to Part 2

We succinctly recall here that the hydrostructural characterization of soil consists of measuring the characteristic hydric properties of the pedostructure, which is to say of the "soil medium", such that we can sample from it a representative volume of about 100 cm^3 within the soil horizon (with homogeneous structure). The pedostructure is the natural physical medium, composed of soil minerals (clay, silt, sand), organized into a dynamic (non-fixed) structure in equilibrium with the water and air that are found inside and which have the possibility of moving relatively to it when submitted to a potential difference.

The pedostructure's hydric properties depend on the thermodynamic interaction between the water and the surface charges of minerals composing the non-rigid soil structure. They are inferred from the pedostructure's four main characteristics, which can be measured precisely in the laboratory:

– *the soil water retention curve (*or the soil water suction): the soil–water tension in absolute value, expressed by suction pressure (hPa) or soil–water retention: $h = f(W)$;

– *the soil shrinkage curve:* specific volume of the pedostructure according to the water content: $\bar{V} = f(W)$;

– *the unsaturated hydric conductivity:* hydric conductivity of the pedostructure, according to its water content: $k_{ps} = f(W)$;

– *the pedostructure swelling curve*, $\overline{V}=f(t)$, by water absorption in the plasma micro-porosity, governed by the equation: $dW_{mi}/dt = k_{mia}(h_{mi}\text{-}h_{ma})$, where k_{mia} is a constant of proportionality to determine, and $(h_{mi}\text{-}h_{ma})$ difference of suction between the water on the interior and exterior of the primary aggregates.

It is necessary to be acquainted with these four characteristics in order to model in detail the pedon's hydrostructural functioning at each level of depth, as done in the "soil–water" Kamel® model [BRA 14a]. This particularly simulates water and air circulation in the different porous systems of the soil subjected to weather conditions such as rain, evaporation (on the surface) and evapotranspiration (at a depth, via the plants).

In the existing soil–water models, all based on the REV (Representative Elementary Volume), only two of these curves are considered in the modeling of water transfer in the soil: the water retention curve $h(W)$ and hydric conductivity $k(W)$. These are the two hydric functions characteristic of the soil that are necessary in order to calculate the water content variations at different depths of the soil with the help of the Richards transfer equation:

$$\frac{d\theta}{dt} = -\frac{\partial(k(dh/dz-1))}{\partial z} \qquad [7.1]$$

where θ is the volumetric water content (m^3/m^3), k is the unsaturated hydric conductivity (m/s) and dh/dx is the suction gradient with elevation z.

In this equation, neither the shrinkage curve $\bar{V}(W)$ nor the swelling speed of primary aggregates appears. As we saw in Part 1, these last two characteristics cannot be integrated in these models based on the REV concept, which results in the non-possibility to physically model the different phenomena linked to the volume variations of the internal organizations with water content, notably the porous volume $\overline{Vp_{mi}}$ of the clay-plasmic (swelling–shrinkage) phase. As example of an important phenomenon, which is impossible to physically model, there is the opening and closing of fissures and cracks with the variation of water content in the soil; or even, the hysteresis of the $h(W)$ function between the humectation and desiccation caused by the water absorption speed by the plasma micro-porosity varying with its filling rate [BRA 06].

On the contrary, the Kamel® model's use of the SREV concept opens it up to these possibilities by allowing for the consideration of the two remaining characteristic curves: $\bar{V}(W)$ and $\bar{V}(t)$, and thus to control the (hydro-) functional volumes of the hierarchized soil organization: structural and functional volumes of the liquid and gas phases in the hierarchized soil organization defined in Part 1 (Table 3.3).

In the Kamel® model, the earlier Richards equation [7.1] was rewritten according to the new paradigm as:

$$dW/dt = -\rho_w \bar{V}\, d\left[k_{ps}(dh/dz - 1)\right]/dz \qquad [7.2]$$

where k_{ps} is the pedostructure hydric conductivity, *z* is the elevation (positive to the top) and *dh*/*dz* is the suction gradient and where *W* and $\bar{V}$ are the two state variables of the soil medium. We remark in passing that due to its structure, its equations and parameters, its inputs and outputs, its interfaces with the other disciplines of environmental sciences, the Kamel® model entirely represents the new paradigm of hydrostructural pedology.

The problem that presents itself when adopting the new paradigm to physically characterize the soil is, therefore, the same problem that presents itself when using the Kamel® model in agro-environmental sciences, that is: how does one obtain the basic information (measuring and determining the parameters of the four characteristic curves) of the soil's hydrostructural functioning that are inputs of the Kamel® model? It is necessary to define not only the methodology for measuring the four characteristic curves but also the method for determining their respective parameters

In regard to the methods for measuring the characteristic curves, they have been outlined in the articles of Braudeau *et al.* [BRA 99] (measuring the shrinkage curve with the retractometer), Braudeau and Mohtar [BRA 06b] (measuring the swelling curve) and Assi *et al.* [ASS 14] for the simultaneous measurement of the shrinkage curve and the water retention using the new TypoSoil[1] tool. However, with regard to the methods to

1 The user manual can be accessed (in OpenAccess) online [BRA 13].

determine the parameters of the characteristic curves[2], no article has yet been outlined. They all come from modeling the theoretical curve on the Excel spreadsheet and using Excel's Solver tool to adjust the theoretical curve to the measured curve.

This is why Part 2 of this book is dedicated to describing the organized Excel sheets for processing the measurement data of the two characteristic curves of soil humidity, those that govern the soil's thermodynamic and hydrostructural equilibrium: the shrinkage and water retention curves. It consists of extracting from them the pedostructure hydrostructural parameters, characteristic to its functioning with water.

Following a succinct theoretical reminder of the equations used, we will consecutively describe the following cases:

1) The water retention curve $h(W)$ in the two measurement cases: using the tensiometer and the porous plate press (curves of the pF);

2) The shrinkage curve $V(W)$, equally in the two measurement cases: alone (measured by the retractometer) and accompanied by the water retention curve (measured by TypoSoil);

3) The hydric conductivity curve $k_{ps}(W)$ measured with the aid of a device with two mounted tensiometers in the laboratory [BRA 07].

The terminology such as the names of the variables, parameters, units as well as the equations used are the same as those used in Part 1, but also in the reference article of the Kamel® model [BRA 14b].

2 The fourth characteristic curve, which will not be presented is the swelling curve in function of time. It can be obtained by measuring the swelling of an initially dry bed of aggregates that have been immersed in water [BRA 06b].

8

Theoretical Recall

8.1. Pinpointing the problem

It is only very recently that the thermodynamic formulation for the soil–water retention curve "WRC" has been established in accordance with the two types of internal and external water contents (W_{mi} and W_{ma}) of the pedostructure primary aggregates [BRA 14c]. Until that point, this curve could only be represented by empirical equations (mathematical equations fitted to the shape of the curve through empirical parameters), or semi-empirical equations, formulated based on assumptions on the morphology of the soil medium structure and the form of interaction of this morphology with water. It was the same for the soil shrinkage curve "ShC" and the soil swelling curve "SwC", for which the equations did not involve the thermodynamics of the equilibrium interaction between water and the soil structure.

Yet today, following the thermodynamic formulation of the soil–water retention curve "WRC" $h(W)$, we know that it is in fact the WRC alone, and not the soil shrinkage curve, which allows us to determine the parameters of the equilibrium curves for the two types of pedostructure water W_{mi} and W_{ma} (and thus also $W_{ip,}$ when it appears). These pedostructure water contents maintain a state of quasi-equilibrium following a slow variation of W by evaporating the soil water. It is this relative variation of these two types of water in a state of equilibrium that configures the soil shrinkage curve. Therefore, it becomes necessary to measure the two curves, shrinkage and retention, simultaneously using the same sample, so that the curve pair's $[(W_{mi}(W), W_{ma}(W)]^{eq}$ parameters (E/A and E_{ma}/A), that we can determine using the retention curve alone, are also those used in the shrinkage curve. Herein lies the real usefulness of the TypoSoil tool.

Before presenting the new principles for determining the characteristic parameters of the pedostructure based on these two new scientific developments: 1) the commercial accessibility of TypoSoil and 2) the thermodynamic formulation of the soil retention curve, we will first succinctly present the approach used to characterize the pedostructure by the shrinkage curve alone, measured by the retractometer, following the Braudeau *et al.* [BRA 04, BRA 05] model.

8.2. Modeling micro- and macro-water types by the shrinkage curve

The inherent problem with "SSM or exponential" semi-empirical models of the soil shrinkage curve, as proposed by Braudeau *et al.* [BRA 99, BRA 04], is the fact that the four types of water considered responsible for the different phases of the shrinkage curve, classically presented with a sigmoidal form (Figure 8.1), are only identifiable on the shrinkage curve of this sigmoidal type. However, this type of curve can only be found in finely aggregated and well-developed soil, where the primary aggregates are well differentiated. This is a structure which highly depends on the type of clay and, for the surface horizon, on its association with organic matter.

In the case of a sigmoidal shape of the shrinkage curve, the semi-empirical models proposed by Braudeau *et al.* [BRA 04] for shrinkage curves measured in continuum with the help of the retractometer are efficient and determine the hydrostructural characteristics in a precise and reproducible manner [BRA 05]. Numerous soil samples from Tunisia and Martinique were analyzed using this method as a part of the development of SIRS-Soil [BRA 01, BRA 07]. The principal hypothesis in these models is that the basic equation of the shrinkage curve can be written in a linear manner, such as:

$$d\bar{V} = K_{re}dw_{re} + K_{bs}dw_{bs} + K_{st}dw_{st} + K_{ip}dw_{ip} \quad [8.1]$$

where, K_{re}, K_{bs}, K_{st}, and K_{ip} are the slopes at the inflection points of the measured shrinkage curve at the basic, structural and interpedal linear shrinkage phases, respectively, in [$\mathrm{dm^3 kg_{water}^{-1}}$], and w_{re}, w_{bs}, w_{st}, and w_{ip} are the water pools associated with the linear shrinkage phases of the pedostructure, in [$\mathrm{kg_{water} kg_{soil}^{-1}}$] (Figure 8.1); $\overline{V}$ is the specific volume of the pedostructure, in [$\mathrm{dm^3 kg_{soil}^{-1}}$]. The distinction between macro and micro was clearly defined and determined by the so called principal shrinkage phase (basic shrinkage phase), between the points C and B (corresponding to pF3 and pF4.2, respectively), where the "macro water" W_{ma}, consisting of W_{st} and W_{ip}, has completely disappeared and no longer remains in the sample, and where what remains in the sample is only the micro water, W_{mi}, consisting of W_{bs} and W_{re}, which in turn does not begin to diminish before the air-entry point (point B in the curve) when the air begins to enter the primary peds.

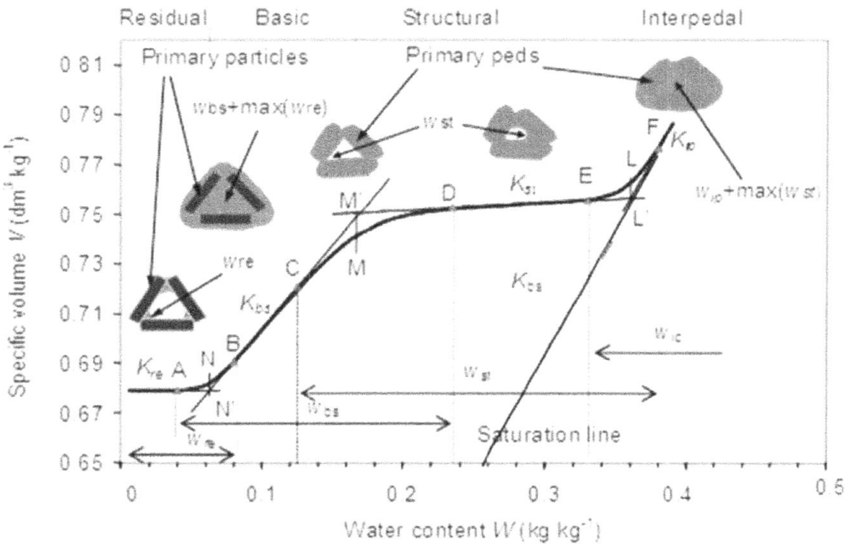

Figure 8.1. *A typical soil shrinkage curve of a well-developed and a fine-aggregate soil structure (pedostructure shrinkage curve). The different configuration of water and air distribution in both the inter- and intra-aggregates' spaces of pedostructure, in relation to the different shrinkage phases of the shrinkage curve.*

The problem with this scenario is that it is only valid when there is no overlapping of the water types' action areas, when the transition points appear in order and are apparent on the shrinkage curve. These therefore have a sigmoidal form, like in Figure 8.1, and are typical of a well-developed structure with fine aggregates.

However, a large number of soil types do not display this clear-cut distinction between the two porous systems, and the shrinkage curve is therefore a uniform curve where the different points of the transition phases do not appear in such a clear manner. Such curves can still be modeled using the soil hydrostructural characterization, but in this case, only four parameters, V_0, W_A, W_B, and the slope at W_B (K_{bs}) (Figure 8.1) are determinable on the shrinkage curve of these types of soil. Therefore, we have tried to build on their retention curves in order to determine their other hydrostructural characteristics [SAL 12]. The equations that were used did however remain semi-empirical, and the ambiguity of certain parameters (especially in the water retention curve) was not alleviated.

With the thermodynamic formulation of the retention curve, taking into account the very real existence[1] of the two aqueous phases called micro and macro (corresponding to W_{mi} and W_{ma}), the distribution of water in these two phases is completely determined by equilibrium at each W value of the pedostructure. In the case of slow evaporation of the soil water tension, the retention curve measured with the tensiometer can be considered as a series of equilibrium states, where, for each W value, the tension measured h is equal to the water tension at equilibrium in the two porous micro and macro systems ($h = h_{mi} = h_{ma}$). The water distribution curves of W_{mi} and W_{ma} (and potentially W_{ip}) are determined exactly by the retention curve and not by the shrinkage curve as we had initially thought. It is in fact this particular distribution of the two types of water, internal and external to the primary aggregates, according to (W) which imposes the form of the pedostructure state equations: $h(W)$, $\bar{V}(W)$, and certainly also $\bar{S}(W)$.

Therefore, the pedostructure hydrostructural characterization will begin by processing the retention curve (measured with the tensiometer or the porous plate-press) as discussed below. This will supply the parameter values for the equilibrium equation of the two types of water W_{mi} and W_{ma}. These values are necessary for processing the shrinkage curve in order to determine its own parameters.

1 For they are separated by a measurable potential difference: $A = E_{ma}/W_{masat} - E_{mi}/W_{misat} \neq 0$ represents the potential difference between the two water phases at the pedostructure saturated state (See Part 1).

8.3. New principle for determining the micro-and macro-water types using the retention curve

8.3.1. *Micro/macro thermodynamic and hydrostructural equilibrium*

The thermodynamic formulation for the retention curve draws on the conceptual model of the organization and distribution of water in the pedostructure that we can describe in the following way: at each state of equilibrium, in terms of W (water content in the pedostructure), the micro aqueous phase (W_{mi}) (water from the clay-humic plasma) and the macro aqueous phase (W_{ma}), which surrounds it all around the medium, are two water phases in the hierarchical organization of the pedostructure which have distinct hydrostructural potential[2]. In this organization, we can say that micro water fills the inter-particle space of the clay plasma, without contact with air. Macro water is the water which occupies the inter-aggregate space, creating an equi-potential interface between the surface of aggregates and the air. These two aqueous phases, of which the quantities W_{mi} and W_{ma} are well defined and determined by the structural arrangement of the soil's solid particles, must necessarily be represented in the thermodynamic soil–water equilibrium equations; and primarily, in the total exact differential of Gibbs' free energy for the liquid phase of the pedostructure ($d\bar{G}_w$), which is written (Part 1, section 6.1.5) as:

$$d\bar{G}_w = -\bar{S}_w dT + \bar{V}_w dP + \mu_{mi} dW_{mi} + \mu_{ma} dW_{ma} = 0, \qquad [8.2]$$

in which the particular feature is that $d\bar{G}_w$ is always equal to 0 at equilibrium, regardless of the pedostructure water content W.

Thus, in the case where the shrinkage and retention curves are measured together using the same sample thanks to the TypoSoil® apparatus, it becomes possible to use the characteristic parameters of the retention curve to originate those of the soil shrinkage curve. The characteristic parameters of WRC can be obtained by curve adjustment, which determines variation curves in terms of W of the two water types (W_{mi} and W_{ma}). Then, by using these parameters, and by fitting the theoretical equation of the shrinkage curve to the measured curve, the remaining characteristic parameters of ShC can be determined. Now, even if the structure in aggregates is not as clear as in the case of the sigmoidal curves, the water organization (distribution)

2 See the previous footnote.

with the solid particles nevertheless remains the same: a two-phase water distribution with an internal water (W_{mi}) surrounded by an external water (W_{ma}) layer, which is the one making the interface with air.

Below, we reproduce the basic equilibrium equations of the two water types and the hydrostructural functioning of the pedostructure derived therefrom. These equations will be used in the Excel spreadsheets to determine the required hydrostructural parameters. We will discuss how the parameters of the soil shrinkage curve can be obtained in each case by using the measurements of the retention curve and the precise determination of its characteristic parameters.

8.3.2. *Equations for the retention curve*

The following equations are obtained from section 3.3.2 of Part 1, to which we may refer for their definitions and set-up. Thus, the equations for the pressure retention of water (or water suction in kPa): in the two porous spaces of different water contents W_{mi} and W_{ma} can be noted as h_{mi} and h_{ma}, respectively, and can be written as:

$$h_{mi} = \rho_w \bar{E}_{mi}(1/W_{mi} - 1/W_{miSat}) \text{ and}$$
$$h_{ma} = \rho_w \bar{E}_{ma}(1/W_{ma} - 1/W_{maSat}). \qquad [8.3]$$

At the point of equilibrium, the suction measured by the tensiometer, h, corresponds to both potentials, such that: $h = h_{mi} = h_{ma}$ and implies a division of W into W_{ma}^{eq} and W_{mi}^{eq}. These water contents at equilibrium are the solutions to a quadratic equation, developed on the basis of the equality of equations 8.3. They are expressed as:

$$W_{ma}^{eq}(W) = \frac{1}{2}\left(W + \frac{\bar{E}}{A}\right) + \frac{1}{2}\sqrt{\left[\left(W + \frac{\bar{E}}{A}\right)^2 - \left(4\frac{\bar{E}_{ma}}{A}W\right)\right]} \qquad [8.4a]$$

and

$$W_{mi}^{eq}(W) = W - W_{ma}^{eq} =$$
$$\frac{1}{2}\left(W - \frac{\bar{E}}{A}\right) - \frac{1}{2}\sqrt{\left[\left(W + \frac{\bar{E}}{A}\right)^2 - \left(4\frac{\bar{E}_{ma}}{A}W\right)\right]}, \qquad [8.4b]$$

where A is a constant representing the difference in chemical potentials between the two types of water at saturation:

$$A = -\left(\mu_{maSat} - \mu_{miSat}\right) = \frac{\overline{E}_{ma}}{W_{maSat}} - \frac{\overline{E}_{mi}}{W_{miSat}} \qquad [8.5]$$

$$\overline{E} = \overline{E}_{mi} + \overline{E}_{ma}$$

where W_{miSat} and W_{maSat} represent the micro- and macro-water types at saturation, such as:

$$W_{Sat} = W_{miSat} + W_{maSat} \qquad [8.6]$$

Equations [8.4a] and [8.4b] determine the division of the micro and macro water in the pedostructure at equilibrium for any amount of water content W.

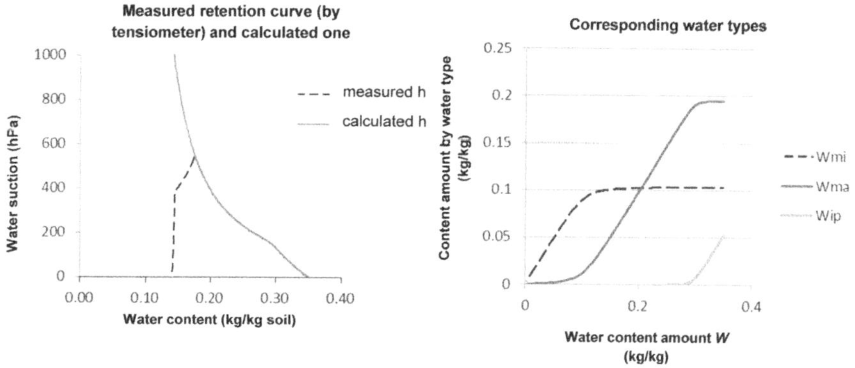

Figure 8.2. *On the left, the retention curve is measured with the tensiometer (in black) and modeled (in red). On the right, the curves of the different water types corresponding to the retention curve. For a color version of the figure, see www.iste.co.uk/braudeau/hydro.zip*

In certain retention curves (for example in Figure 8.2), two ascending phases of suction are observed to be separated by a bulge at around 100–300 hPa, which appear to be a continual transition between the two phases. In all of the cases observed, the first retention phase, going from saturation ($h = 0$) to the bulge, corresponds in a very clear manner to a shrinkage phase called "interpedal saturation shrinkage phase", which appears at the same water content on the associated shrinkage curve. This shrinkage phase has a slope of 1 ($K_{ip} = 1$) parallel to the saturation line (Figure 8.1). This bulging, due to the presence of W_{ip}, of which the absorption in the inter-aggregates macro-porosity, in addition to W_{ma}, is compensated by spacing the aggregates between them and therefore the sample swelling.

The global water content, including this water W_{ip} in surplus, is noted as W' and is such that:

$$W' = W + W_{ip} = W_{ma} + W_{mi} + W_{ip}. \qquad [8.7]$$

It is important to note that equations [8.4a] and [8.4b] are written with the variable W and not W'.

The pressure retention h_{ip} of the water type (W_{ip}) depends, like W_{ma}, on the surface charges of the primary aggregates ($-\bar{E}_{ma}$). It therefore has an analogous expression of h_{ma} and adds onto this last one in a way, such as:

$$h = h_{mi} + h_{ip} = h_{ma} + h_{ip} \qquad [8.8]$$

and

$$h_{ip} = \rho_w E_{ma}/(W_{ip^\circ} + W_{ip}) - \rho_w E_{ma}/(W_{ip^\circ} + W_{ipSat}), \qquad [8.9]$$

where W_{ip} is the water content which has been defined by interpreting the shrinkage curve (equation [8.1]) and is expressed as:

$$W_{ip} = (1/k_L)\text{Ln}(1 + \exp[k_L(W - W_L)]), \qquad [8.10]$$

in which k_L and W_L are the two parameters of this shrinkage phase (from W_F to W_E on the shrinkage curve, Figure 8.1). W_{ip° is a characteristic parameter of the equation of h_{ip} (equation [8.9]) that we obtain by adjusting the theoretical curve to the measured curve. Its approximate value is calculated according to equation [8.9] assuming that $W_{ip} = 0$ for $h_{ip} = h_{ip^\circ}$, the value of suction at the bulging, the beginning of the second shrinkage phase. The estimated value W_{ip° is therefore a solution to the equation of the 2nd degree polynomial as follows:

$$h_{ip^\circ} W_{ip^\circ}{}^2 + h_{ip^\circ} W_{ipSat} W_{ip^\circ} - \rho_w E_{ma} W_{ipSat} = 0 \qquad [8.11]$$

Thus, we obtain:

$$W_{ip°} = \frac{1}{2}\left(-W_{ipSat} + \sqrt{W_{ipSat}^{2} + 4\rho_w E_{ma} W_{ipSat}/h_{ip°}}\right), \quad [8.12]$$

where we also have $W_{ipSat} \approx (W'_{Sat} - W_L)$ (according to equation [8.10]).

In summary, the parameters of the water retention (tensiometric) curve are:

– four parameters in case of the absence of an interpedal phase (ip) (in this case $W_L = W'_{sat}$): E_{mi}, E_{ma}, W_{miSat}, and W_{maSat}.

– seven parameters in case of the presence of an interpedal phase (ip) (in this case, $W'_{sat} - W_L >> 0$): E_{mi}, E_{ma}, W_{miSat}, W_{maSat}, k_L, $W_{ip°}$, and W'_{Sat}.

We note that in this last case:

$$W_{maSat} = (W_L - W_{miSat}) \text{ and } W_{ipSat} = (W'_{Sat} - W_L)). \quad [8.13]$$

However, in either case, we prefer to adjust the theoretical curve to the measured curve by directly using the parameters of the equations [8.4a, b]: $D = -E_{ma}/A$ and $F = -E/A$. These parameters are the thermodynamic equation parameters of the two water types at equilibrium (W_{mi}^{eq} and W_{ma}^{eq}) which are common to all the pedostructure state equations (i.e. equations of the hydrostructural functioning: the pedostructure water retention curve and the shrinkage curve). These two parameters, D and F, will therefore be used in place of two of the parameters listed above (E_{mi}, E_{ma}, W_{miSat}, and W_{maSat}), which will be calculated based on the first two parameters (D and F) by the relationships as discussed below.

First of all, we find that knowing the value of D and F imposes a unique relationship between W_{miSat} and W_L:

By definition, $D = -E_{ma}/A$, $F = -E/A$ and according to [8.5]:

$$A = \frac{\bar{E}_{ma}}{W_{maSat}} - \frac{\bar{E}_{mi}}{W_{miSat}} = \frac{\bar{E}_{ma}}{(W_L - W_{miSat})} - \frac{\bar{E}_{mi}}{W_{miSat}} \quad [8.14]$$

$$A = \frac{\bar{E}_{ma} W_{miSat} - \bar{E}_{mi}(W_L - W_{miSat})}{W_L W_{miSat} - W_{miSat}^{2}}$$

By dividing each member by A and by introducing the variables F and D into the equation, the following polynomial equation of the 2nd degree is obtained:

$$W_{miSat}^{\ 2} - W_{miSat}\left(W_L + F\right) + W_L(F - D). \quad [8.15]$$

The solution of this equation, which fits for W_{miSat}, is:

$$W_{miSat} = \left[\left(W_L + F\right) - \sqrt{\left(W_L + F\right)^2 - 4W_L(F - D)}\right]/2. \quad [8.16]$$

The equation of W_{miSat} is valid in the case of either the presence or absence of the interpedal phase, respectively corresponding to the case where ($W_L \ll W_{sat}$) and ($W_L = W_{sat}$); W_{miSat} can therefore be calculated from equation [8.16] by knowing the values of D, F and W_L that are obtained by adjustment. $W_{mi\,Sat}$ will be one of the two parameters to be replaced by F and D.

In regards to the 2nd parameter to be replaced, we can ascertain that because $F/D = E/E_{ma} = 1 + E_{mi}/E_{ma}$, E_{ma} can be calculated based on the parameters D, F and E_{mi} by the equation:

$$E_{ma} = E_{mi}/(F/D - 1). \quad [8.17]$$

Consequently, the chosen parameters to be determined by adjusting the theoretical retention curve to the measured curve, with the help of Excel's Solver, are:

– in the case where $W_L = W_{Sat}$ (no ip phase): $F, D,\ E_{mi},$ and W_{Sat};

– in the case where $W_L \ll W_{Sat}$: $F, D,\ E_{mi},\ W_{Sat},\ k_L,\ W_L,$ and $W_{ip°}$.

8.3.3. *Equations for the pF curve*

The equation that needs to be known in order to determine the adjustment parameters of the pF curve is the following equation [8.18], which transforms, on the ordinate axis, the air pressure applied to the sample in Richards' porous plate-press, in the water retention pressure of the pedostructure given by Braudeau *et al*. [BRA 14d]:

$$h_i = -\rho_w \Delta\mu_w = \rho_w\, RT/M_w\, \mathrm{Ln}\left(\left(P_a + \Pi_i\right)/P_a\right), \quad [8.18]$$

where R is the constant of perfect gases, T is the absolute temperature, P_a is the atmospheric pressure of the measurement conditions, and Π_i is the air pressure added in the press corresponding to the measurement (i).

Practically, this transforms the air pressure (Π_i) corresponding to the measurement points into the soil water retention h_i by using the equation $T = 294$ K (21°C):

$$h = 137.72 \ln\left(\frac{\Pi}{100} + 1\right), \quad [8.19]$$

where h and Π are expressed in kPa [BRA 14d].

An example of a retention curve derived from the pFs curve is given in Figure 8.3. The curve formed by the measurement points $[h_i, W_i]$ obtained under the pressure Π_i is therefore conceptually identical to the retention curve measured by the tensiometer. Determining the parameters of the curve by adjustment will require a method analogous to that presented in the previous section, however it must be acknowledged that the presence of a potential phase (ip) will only be a rough guess, and that its detection will only be possible if the measurements are of a sufficient number near the saturation.

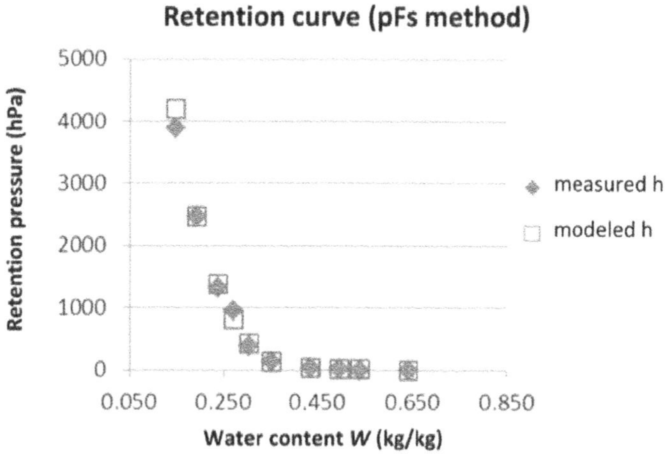

Figure 8.3. *Example of a soil water retention curve derived from the pFs curve measured under air pressure in the porous pressure plate (Richards' apparatus)*

8.3.4. *Equations for the shrinkage curve*

8.3.4.1. *The case for non-sigmoidal shrinkage curves*

We have seen that the case of the shrinkage curve in a very clear sigmoidal form is a particular case of the shrinkage curve, where the water types (w_{re}, w_{bs}, w_{st}, w_{ip}) responsible for the curve's linear parts successively disappear from the sample, only overlapping in the curved parts (Figure 8.1). If the transition points of the phase are not clearly identifiable on the curve, as in the two examples given in Figure 8.4, that means that the water types are overlapping all along the curve. In particular, W_{ma} (identified in the new equations of thermodynamic origin to be w_{st}) is no longer of a negligible quantity along the course of the so-called "basic" shrinkage phase, nor near the air entry point which marks the beginning of the residual shrinkage phase. The slope K_{bs} of the shrinkage curve at the air entry point is therefore necessarily inferior to the coefficient of w_{bs} in equation [8.1], a coefficient that we can no longer identify with the curve slope and that we name $K_{bs}^{'}$:

$$K_{bs} = d\bar{V}/dW = d\bar{V}/d(w_{bs} + w_{st}) < K_{bs}^{'} = d\bar{V}/d(w_{bs}). \qquad [8.20]$$

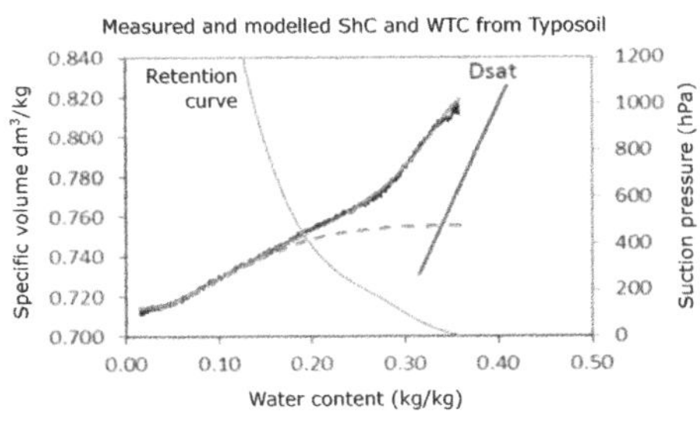

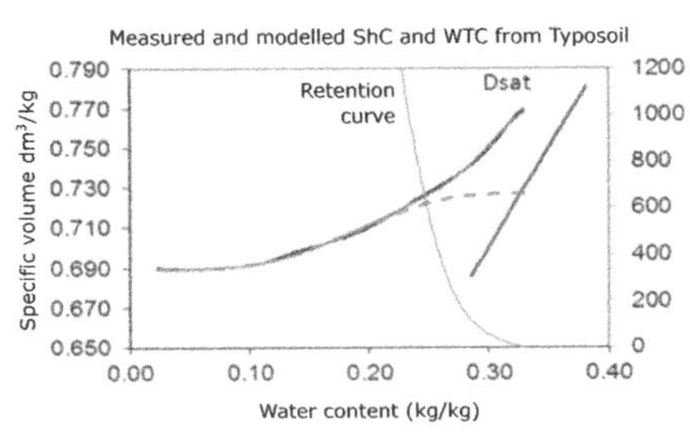

Figure 8.4. *Two examples of non-sigmoidal shrinkage curves (red and blue) linked to their retention curve in green (Braudeau et al. [BRA 14c]). For a color version of the figure, see www.iste.co.uk/braudeau/hydro.zip*

When adjusting the shrinkage curve, the slope of the principle (basic) shrinkage phase cannot be a fixed parameter (geometrically obtained by measuring the curve's slope), but is one more parameter to be determined in the adjustment. Extracting the parameters by adjustment is therefore not possible if we refer to the shrinkage curve alone; there are too many possible solutions. On the other hand, if we use the parameters obtained from the retention curve of the same sample, which are: D, F, W_{sat}, and potentially W_L and K_L, the remaining parameters of the shrinkage curve become effectively determinable by adjusting the theoretical curve to the measured curve.

The equation for the shrinkage curve that is finally retained is:

$$d\bar{V} = K'_{bs}d(W_{mi} - w_{re}) + K'_{st}dw_{st} + K_{ip}dW_{ip}, \qquad [8.21]$$

which recalls the earlier equation [8.1], but in which K_{bs}, the curve's slope at the air-entry point, will have been replaced in the equation by K'_{bs}, a coefficient of defined structure such that $K'_{bs} = d\bar{V} / d(w_{bs})$, and where $K_{re} = 0$ has disappeared as it no longer has a place to be[3]. We also note that K_{ip} is generally equal to 1, and K'_{st} to zero. The integration of equation [8.21] starting from $W = 0$ gives:

$$\bar{V} = \bar{V}_o + Kbs(Wmi\text{ -}wre) + K'_{st}W_{ma} + K_{ip}W_{ip}, \qquad [8.22]$$

where $\bar{V}_o$ is the specific volume of the sample at the shrinkage limit, $W_{mi} = w_{bs} + w_{re}$, $W_{ma} = w_{st}$, and $W_{ip} = W - W_{ma} - W_{mi}$.

The content of both residual and interpedal water, w_{re} and W_{ip}, are calculated by the following equations given in Braudeau *et al.* [BRA 14a]:

$$w_{re}^{eq} = W_{mi}^{eq} - \left(1/k_N \right) Log\left[1 + \exp\exp\left(k_N\left(W_{mi}^{eq} - W_{miN}\right)\right)\right] \qquad [8.23]$$

and

$$W_{ip} = \left(1/k_L \right) Log\left[1 + \exp\exp\left(k_L\left(W - W_L\right)\right)\right], \qquad [8.24]$$

3 In certain cases, the residual phase appears with a slope Kre > 0 which indicates the presence of a porous subsystem of mineralogical (e.g. volcanic ash) or organic material.

where W_{miN}, k_N, k_L, and W_L are the parameters of equations [8.23] and [8.24] to determine, as well as the other characteristic parameters of the shrinkage curve (equation [8.22]), namely: $\bar{V}_o$, K_{bs}', K_{st}', K_{ip} and the parameters F and D from the equations of W_{mi}^{eq} and W_{ma}^{eq}.

Therefore, if the retention curve is measured simultaneously with the shrinkage curve, using the TypoSoil tool, the parameters $D = -E_{ma}/A$ and $F = -E/A$, as well as k_L and W_L when the (ip) phase is present, are perfectly determined by previously adjusting the retention curve. The shrinkage curve can therefore be adjusted by varying only its own parameters: K_{bs}' k_N, and W_{miN}; and potentially K_{st}' and K_{ip}.

In summary, to adjust the shrinkage curve, we first begin by adjusting the retention curve which will provide the shrinkage curve with imperative characteristic parameters: D and F, and, potentially, k_L and W_L.Then after having fixed the parameters ($\bar{V}_o$ and W_{miN}), which can be read on the shrinkage curve, we can determine the remaining parameters (K_{bs}', k_N, W_{miN}) by adjusting the theoretical curve to the measured curve. It may be necessary to re-specify k_L and W_L. K_{st} and K_{ip} are generally fixed to 0 and 1, respectively.

8.3.4.2. *The case for sigmoidal shrinkage curves*

When the curve has a sigmoidal shape and its transition points of phases (A, B, C, etc., Figure 8.1) can be clearly identified on the curve, it is possible to calculate the two water types W_{mi} and W_{ma} in terms of W according to the shrinkage curve alone (without needing to know the retention curve). The curve adjustment is done by fixing K_{bs}', the parameter of equation [8.22], to the value of the slope of basic shrinkage (K_{bs}) as read on the curve. In doing so, ($K_{bs}' = K_{bs}$), we consider that the sample has a fine aggregate structure that is well developed. According to equation [8.21], $W = W_{mi}$ along this slope segment: at point C of the curve, W_{ma} becomes negligible and the remaining water in the sample is no longer found anywhere except in the plasmic porosity of the primary aggregates (i.e. W_{mi}). The water types do not overlap and we can use the relationships given in the Structural Shrinkage Model [BRA 93, BOI 06] or the EXP model [BRA 04, BRA 05] to interpret the shrinkage curve in terms of water and air volume in the two micro- and macro-porous systems, at the different transition points of the shrinkage curve.

Thus, K_{bs}' and $\bar{V}_o$ being fixed to their value read on the (sigmoidal) shrinkage curve, the remaining parameters to be determined using the adjustment method are: W_{miN}, k_N, D (=− E_{ma}/A), F (=−E/A) as well as k_L and W_L in the case of the (ip) phase being present (K_{ip} = 1).

In opposition to the previous case of non sigmoidal curves, the last four parameters, which are also those for the equations of the water types W_{mi}, W_{ma}, and W_{ip}, can be determined by the shrinkage curve alone, due to the fact that it has a clear sigmoidal form, signifying a developed aggregated structure.

8.3.5. *Equations for hydric conductivity*

In order to establish the hydric conductivity equation and to be able to determine its parameters, we must refer to the diagram showing the measuring device in Figure 8.5. The sample is a soil cylinder, either disturbed or undisturbed, 5 cm in diameter and 5 cm in height. It is introduced into a hollow cylinder with the same interior dimensions, which has been pierced at the heights of 1 and 2 cm to allow the two mini tensiometers T1 and T2 to pass through. The tubes of the two tensiometers are filled with water and are linked to the pressure sensors located in a tensiometric casing[4], which is itself linked to a computer.

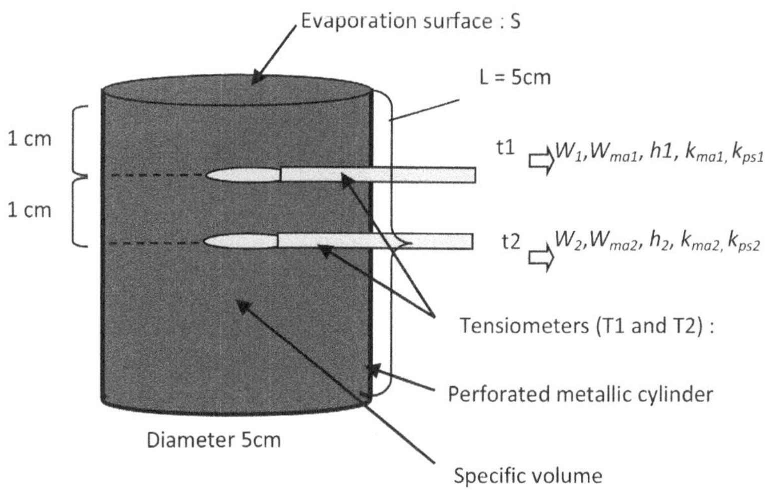

Figure 8.5. *Diagram representing the sample and its conditioning for the hydric conductivity measurement*

4 Part of SDEC commercial material for measuring hydric conductivity using the Wind method.

Calculating the conductivity is based on evaporation which occurs exclusively on the upper face of the sample; a plastic film is thus placed on the lower face and a ring, with a diameter equal to the cylinder's diameter and a few millimeters thick, is placed on the upper face so as to limit lateral evaporation, which is produced by the shrinkage of the sample, thus diminishing its diameter.

The potentials obtained from the two tensiometers are presented under the form of tensions T (expressed in mV). A calibration was necessary to convert these tensions into pressures h (expressed in hPa). The whole device, cylinder and tensiometers, is placed on the scale to measure the weight. The entire device is put into a chamber at 40°C for slow and regular water evaporation through the cylinder's surface (Figure 8.6).

Figure 8.6. *Photo of a laboratory device which allows us to measure the hydric conductivity of the soil [PRA 05]*

This allows us to know the total flux of water leaving the sample F_{tot} and also, after measuring the dry sample mass (M_s) after being oven dried at 105°C at the end of the experience, the total water content W_{tot} of the sample throughout the experience can be determined.

Tension and mass are automatically recorded every 5 min. At the end of the handling, the value series (time and mass) can also be copied from the acquisition computer and entered into an Excel sheet for raw data processing, which will be used as a base for analyzing the results.

This was the procedure used at the pedology laboratory of IRD and PRAM (the Agronomic Research Unit of Martinique) on some twenty samples of hydrostructural characterized soil in Martinique as part of the "SIRS-Soils of Martinique" project [BRA 07]. The data used as an example in Chapter 9 (section 9.3) are obtained from the SIRS-Soil database.

The hypotheses made by calculating the hydric conductivity are as follows:

1) the total water content of the sample, W_{tot}, is identical in value to that which is located on the level of the tensiometer T2. We will therefore assume that the parameters extracted from the tensiometric curve that consists of measuring points $[h_2, W_{tot}]$ are characteristic to the sample;

2) the conceptual model, which underlies the calculation of the hydric conductivity, is represented in Figure 8.7. The variables W_{ma}, V_{ps} and k_{ps} are considered as the variables of the *i* layer's state between the tensiometers T1 and T2. W, W_{ma} *and* V_{ps} are calculated as averages of the values in T1 and T2.

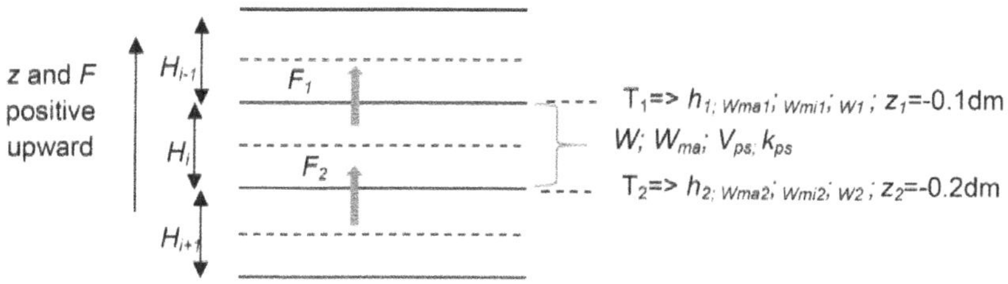

Figure 8.7. *Conceptual model for calculating the hydric conductivity of the layer between tensiometers T1 and T2*

We could now avoid hypothesis 1 by measuring beforehand the hydrostructural characteristics of the sample with the TypoSoil® tool. In this way, we would have the exact local values of W_{ma}, W_{mi}, and W (by using equation [8.3]) at each measurement h_1 and h_2 instead of those calculated based on the parameters extracted from the curve h_2 = f(W_{tot}), which is approximate.

The Richards' equation is written as follows:

$$\frac{dW}{dt} = -\rho_w \bar{V} \frac{\partial}{\partial z}\left(k_{ps}\frac{\partial}{\partial z}(h - z)\right), \qquad [8.25]$$

where z is the elevation (positive upwards), k_{ps} is the pedostructure hydric conductivity, and $\bar{V}$ is the specific volume of the pedostructure. Clarifying the derivative with respect to z, we obtain:

$$\frac{dW}{dt} = -\rho_w \bar{V} \frac{\partial}{\partial z}\left(k_{ps}\frac{\partial(h-z)}{\partial z}\right) = -\rho_w \bar{V}\left\{\left(\frac{\partial(h-z)}{\partial z}\right)\frac{\partial k_{ps}}{\partial z} + k_{ps}\frac{\partial^2 h}{\partial z^2}\right\}. \qquad [8.26]$$

Assuming that the layer of soil between the two tensiometers is quite fine, we can rewrite equation [8.26] under the below form:

$$\frac{dW}{dt} = -\rho_w \bar{V}\left\{\frac{\Delta k_{ps}}{\Delta W_{ma}}\left(\frac{\Delta h - \Delta z}{\Delta z}\right)\frac{\Delta W_{ma}}{\Delta z} + k_{ps}\frac{\Delta(dh/dz)}{\Delta z}\right\}, \qquad [8.27]$$

where Δh, $\Delta z, \Delta k_{ps}$ and ΔW_{ma} represent the value differences of the variables in question between the tensiometers 1 and 2.

The derivative's first term is negative in the measurement conditions (evaporation: $dW_{ma}/dz > 0$, $dk_{ps}/dW_{ma} < 0$ and $dh/dz > 0$), whereas the second term must necessarily be positive in order to counterbalance the first term's contribution to dW/dt, which must necessarily be negative as there is a loss of water from the sample due to evaporation.

With regards to the second term, we can rewrite it in terms of W_{ma} in the following way:

$$k_{ps}\frac{\Delta(dh/dz)}{\Delta z} = k_{ps}\frac{\Delta(dh/dz)}{\Delta W_{ma}}\frac{\Delta W_{ma}}{\Delta z} = k_{ps}\frac{\Delta W_{ma}}{\Delta z}\frac{\Delta}{\Delta W_{ma}}\left(\frac{dh}{dW_{ma}}\frac{dW_{ma}}{dz}\right) \qquad [8.28]$$

$$k_{ps}\frac{\Delta(dh/dz)}{\Delta z} = k_{ps}\left(\frac{\Delta W_{ma}}{\Delta z}\right)^2 \frac{d^2 h}{dW_{ma}{}^2}, \qquad [8.29]$$

Where, according to equation [8.3], the second derivative of h with respect to W_{ma} is:

$$\frac{d^2 h}{dW_{ma}{}^2} = 2\frac{\rho_w \bar{E}_{ma}}{W_{ma}{}^3} \qquad [8.30]$$

In Chapter 9 we will show, with an example of measurement using the device in Figure 8.6, that the two multiplication factors in equation (8.27) of dk_{ps}/dw and k_{ps}, namely:

$$\left(\frac{\Delta h-\Delta z}{\Delta z}\right)\frac{\Delta W_{ma}}{\Delta z} \text{ and } \frac{\Delta(dh/dz)}{\Delta z} = \left(\frac{\Delta W_{ma}}{\Delta z}\right)^2 \frac{d^2h}{dW_{ma}{}^2}, \qquad [8.31]$$

can be calculated very precisely based on the measurement results supplied by the device, and that they are exponential functions of W_{ma} that we found with correlation coefficients exceeding 0.999 (Figure 8.8).

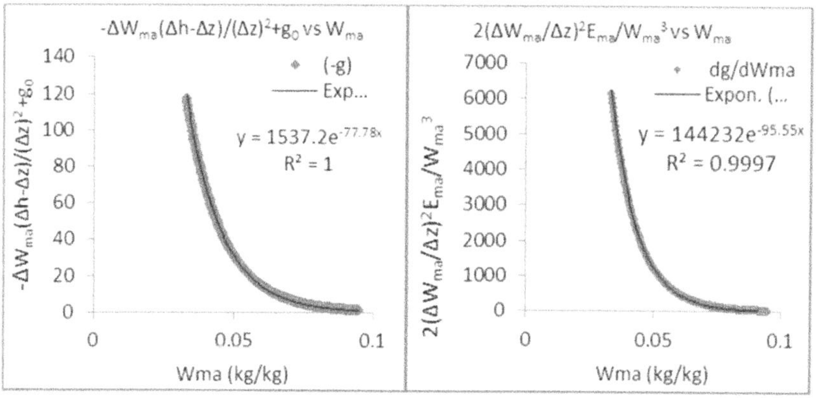

Figure 8.8. *Example of measured curves, according to W_{ma}, of the factors [8.31] of the two terms $\Delta k_{ps}/\Delta W$ and k_{ps} of the Richards' equation [8.25]*

Let us call the first term g, factor of $\Delta k_{ps}/\Delta W_{ma}$ in equation [8.27]. According to the result shown in Figure 8.8, the equation for g (negative) is:

$$g = \left(\frac{\Delta h-\Delta z}{\Delta z}\right)\frac{\Delta W_{ma}}{\Delta z} = -k_1 Exp(\alpha_1 W_{ma}) + g_0. \qquad [8.32]$$

Its derivative in relation to W_{ma} is, in fact, exactly equal to the second term in equation [8.31] as given below:

$$\frac{d}{dW_{ma}}\left[\left(\frac{dh-dz}{dz}\right)\frac{dW_{ma}}{dz}\right] = \frac{dW_{ma}}{dz}\frac{d}{dW_{ma}}\left(\frac{dh}{dW_{ma}}\frac{dW_{ma}}{dz}\right) = \left(\frac{dW_{ma}}{dz}\right)^2\frac{d^2h}{dW_{ma}{}^2} \qquad [8.33]$$

Consequently, for this second term in Figure 8.8, we must find an exponential form as a function of W_{ma}, such as:

$$\frac{\Delta(dh/dz)}{\Delta z} = \frac{dg}{dW_{ma}} = -\alpha_1 k_1 Exp(\alpha_1 W_{ma}). \qquad [8.34]$$

However, on the graphs, we can find that the W_{ma} coefficients: α_1 and α_2 for the two exponential curves are notably different, although they are from the same order; the same can be said for the product $\alpha_1 k_1$, which should be found to be equal to k_2. Despite these gaps, that can be attributed to the hypothesis concerning the water content calculated in T_2 as identified at W_{tot}, we will consider both to be true, the equation [8.32] of g, the first factor, and the equation [8.34] for the factor as the derivative of g in respect to W_{ma}. Equation [8.27] will therefore be written based on g as:

$$dW/dt = -\rho_w \bar{V}\left(\frac{dk_{ps}}{dW_{ma}} g + k_{ps}\frac{dg}{dW_{ma}}\right) = -\rho_w \bar{V}\frac{d(k_{ps}\, g)}{dW_{ma}}. \qquad [8.35]$$

In the example where dW/dt according to W_{ma} is akin to a straight line comprising the parameters a and b (the case of this example), after integration between W_{ma}=0 and W_{ma}, the equation for k_{ps} becomes:

$$k_{ps} = \frac{(aW_{ma}{}^2 + bW_{ma})}{\rho_w \bar{V} g} \approx k_{ps}^0 W_{ma} Exp(\alpha_{ps} W_{ma}), \qquad [8.36]$$

in which the parameters are $\alpha_{ps} = -\alpha_1$ and $k_{ps}^0 = -\,b/(\rho_w \bar{\bar{V}} k_1)$, considering that the term $aW_{ma}{}^2$ is quickly negligible against bW_{ma}, and g_0 is small compared with $k_1\exp(\alpha_1 W_{ma})$, the validity of the measurement range (Δh at least greater than 1 kPa), as we will discuss for the example given in section 9.3.3.

These different equations are used in the Excel sheet dedicated to determining the pedostructure hydric conductivity, which is presented at the end of the following chapter (section 9.3).

9

Methods for Determining the Characteristic Parameters

9.1. Soil water retention curve "WRC"

9.1.1. *Measured using the tensiometer (suction-based method)*

The soil water retention curve (WRC) is a relationship between the water content (W) and the soil–water potential (h). On the Excel spreadsheet dedicated to adjusting the tensiometric curve, the Typosoil measurements for (W) and (h) are input by copying and pasting them into the first two columns of the Excel sheet, as shown in Figure 9.1[1].

There are two types of water retention curves based on the presence or absence of the interpedic saturation phase visible on the measured curve (in black in the diagram). The interpedic phase is due to the interpedal water W_{ip} and occurs near saturation of the sample. Its presence can be seen on the measured curve by a rapid decrease in the water suction after a bulging appearing at a suction called $h_{ip}°$ at around 100–200 hPa, as in the example of Figure 9.1. We recall (section 8.3.2) that this bulging is identified with the shrinkage phase around point L of the soil shrinkage curve (ShC) and that, in this case, $W_L = W_{sat} << W'_{sat}$.

1 All of the Excel spreadsheets included in this chapter can be downloaded from the editor's website.

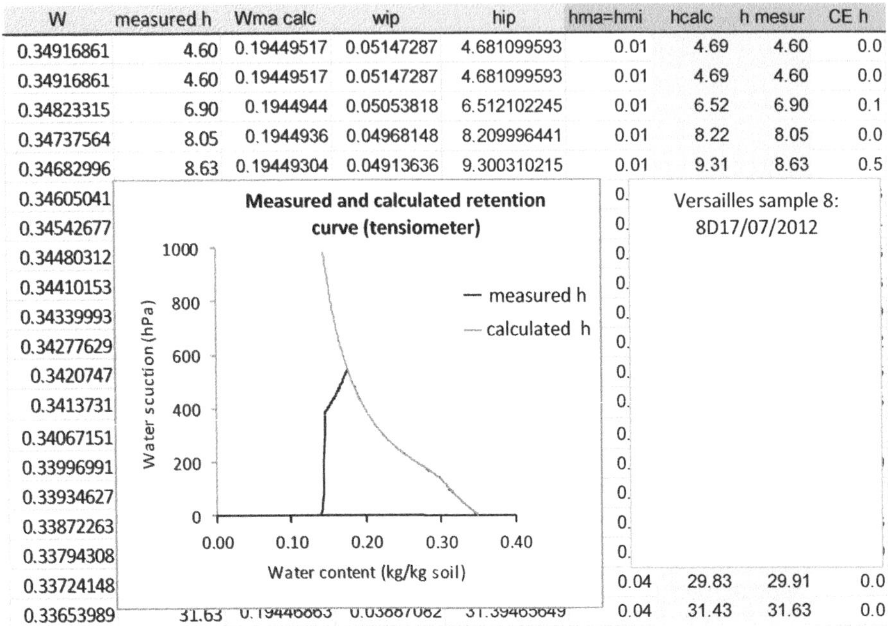

W	measured h	Wma calc	wip	hip	hma=hmi	hcalc	h mesur	CE h
0.34916861	4.60	0.19449517	0.05147287	4.681099593	0.01	4.69	4.60	0.0
0.34916861	4.60	0.19449517	0.05147287	4.681099593	0.01	4.69	4.60	0.0
0.34823315	6.90	0.1944944	0.05053818	6.512102245	0.01	6.52	6.90	0.1
0.34737564	8.05	0.1944936	0.04968148	8.209996441	0.01	8.22	8.05	0.0
0.34682996	8.63	0.19449304	0.04913636	9.300310215	0.01	9.31	8.63	0.5
0.34605041					0.			
0.34542677					0.			
0.34480312					0.			
0.34410153					0.			
0.34339993					0.			
0.34277629					0.			
0.3420747					0.			
0.3413731					0.			
0.34067151					0.			
0.33996991					0.			
0.33934627					0.			
0.33872263					0.			
0.33794308					0.			
0.33724148					0.04	29.83	29.91	0.0
0.33653989	31.63	0.19446663	0.03687062	31.39465649	0.04	31.43	31.63	0.0

Figure 9.1. *Copy of the top left section of the Excel sheet dedicated to adjusting the retention curve measured with the tensiometer. For a color version of the figure, see www.iste.co.uk/braudeau/hydro.zip*

The series of variable columns shown in Figure 9.1 can be grouped into three types: 1) the measured data: columns 1 and 2 are the measured water content (W) and water suction (h); 2) the calculated "modeled" data: columns 3 through 7 are the calculated state variables of the different pedostructure pore systems by using the pedostructure WRC equations (section 8.3.1) and the measured water content and suction. Column 3 is the calculated pedostructure macro-pore water content (W_{ma}). Column 4 is the interpedal water content (w_{ip}). Column 5 is the interpedal water suction, i.e. the water suction due to W_{ip} and associated with the interpedic saturation phase. Column 6 is the calculated suction pressure created by the micro-water in exact opposite to that created by the macro-water ((h_{mi}) and (h_{ma})). Column 7 is the calculated pedostructure water suction, such that: $h_{calc}=h_{mi}+h_{ip}=h_{ma}+h_{ip}$; 3) The modeled and optimized data: column 7 shows the modeled water suction to be compared with the measured water suction in column 8; and column 9 is the square of the differences between the measured and calculated water suction (h), which should be minimized by using Excel solver.

As discussed before, there are generally two types of the pedostructure water retention curves based on the presence or absence of the interpedic saturation phase. Actually, the Excel spreadsheet displays those both options, with or without this interpedic phase (ip), which need to be processed differently.

9.1.1.1. *1st case: without the interpedic (ip) saturation phase: $W_L = W_{Sat}$*

Before using Excel's Solver tool to make the adjustments, we first need to set up the extraction of the parameters by filling in the parameter tables in Figure 9.2 (a) and (b)).

9.1.1.1.1. Information about the adjustment conditions

In Figure 9.2(a) $h_{ip}°$ is put at 0 as there is no (ip) phase. This curve point ($h_{ip}°$, in hPa) is the one that marks the beginning of the fitting range of the theoretical curve to the measured curve.

The "end of validity" case is filled with the suction value from which the curve begins inflecting (between 500 and 900 hPa) before the break of the tensiometer water column. This value can be read by hovering the mouse over the recognized point on the water retention curve.

Coefma is set to 0. This parameter indicates that the option to choose for processing the data is the one for processing the shrinkage and retention curves with no (ip) phase, whether this ($h_{ip}° > 0$) actually exists or not ($h_{ip}° = 0$). This essentially means making the parameters inactive relative to this phase (k_L, W_L and $W_{ip°}$) on the right hand-side of Figure 9.2(b).

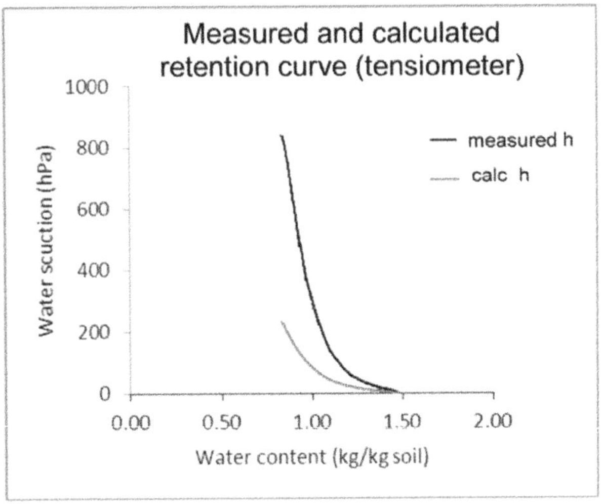

a)

Putting the values onto the graph		
hip° (0 or ...):	0	hPa
End of validity:	730	hPa
Coefma	0	

b)

Target solver		Modeling Interval	
1.00E+07		1	730
Parameters to be determined			
Wsat kg/kg	1.500	**kL**	139.15
(-F)	1.000	**WL**	0.298
(-100*D)	1.000	**Wipo**	0.103
Emi =	100.0		
Cte hip°	0	Wip° sugg.	2.935

Figure 9.2. *Example for setting up and initializing the parameters before the curve adjustment by Excel solver. Case I – No interpedic saturation phase: (a) is to set up the optimization process based on the shape and readings of the WRC: (b) shows the optimization process and the parameters to be determined [W_{sat}, W_{miSat}, W_L, k_L, $W_{ip}°$, E_{ma}, E_{mi}, and $h_{ip}°$]. For a color version of the figure, see www.iste.co.uk/braudeau/hydro.zip*

9.1.1.1.2. Inputting the parameters to be determined

Figure 9.2(b) displays the parameters for the theoretical retention curve. They will be determined by adjusting the retention curve to the measured curve, using Excel's Solver tool. First, an initial value must be attributed. For *Coefma* = 0, the present case, only the left-hand side of the table is involved:

– W'_{sat} is the value of the water content at saturation (h = 0), which can be estimated quite precisely by using the data measurements, but it can be determined even more precisely (for h = 0) by using the Excel solver;

– $F = -E/A$ and $D = -E_{ma}/A$ are the two parameters of the equations for W_{mi} and W_{ma} at equilibrium. ($-F$) is initialized with a value approximatively equal to the water content read on the retention curve at the beginning of the rapid increasing of suction ($-F$ = 1.0 in the example shown in Figure 9.3); and (−100 * D) is generally initialized at 1;

– E_{mi} is the parameter of the equation, $h_{mi}(W_{mi})$; it is initialized with any one value between 40 and 100 J/kg.

Cte $h_{ip}°$ is a constant that represents $h_{ip}°$ and is initialized with the value given in the previous table. In the present case, *Cte* $h_{ip}°$ = 0. This constant is determined when there is an (ip) phase and when it is worked out with Coefma = 0. *Cte* $h_{ip}°$ synthesizes and replaces the parameters of the (ip) phase. Its usefulness will be discussed later in this chapter.

9.1.1.1.3. Using the Excel solver

The solver is then activated, the optimization method chosen is the GRG non-linear method, the object to be minimized (Target Solver) is the sum of the squared derivations (column "CE h", Figure 9.1) and the variable parameters to be determined are the first four parameters on the left-hand side of Figure 9.2(b). We begin with the three parameters ($-F$), (−100 * D) and E_{mi} while maintaining W_{sat} fixed at its estimated value on the curve. Then, while the solution for these three parameters is being found, we perfect the adjustment of by adding W'_{sat} to the three others.

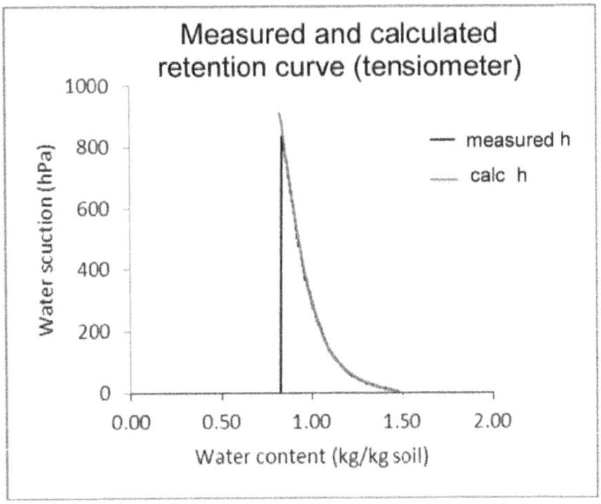

Target solver		Modeling Interval	
7.64E+03	4.9	1	775
Parameters to be determined			
Wsat kg/kg	1.499	**kL**	139.15
(-F)	1.025	**WL**	0.298
(-100*D)	0.691	**Wipo**	0.103
Emi =	372.9		
Cte hip°	0	Wip° sugg.	4.948

Results			
W_{sat} kg/kg	W_{miSat} kg/kg	W_L	W_{ip}°
1.499	1.004	No signific.	No signific.
E_{ma} J/kg	E_{mi} J/kg	k_L	Cte h_{ip}°
2.53	372.87	No signific.	0

Figure 9.3. *Components of the retention curve adjustment sheet. Adjustment results from the case where there is no interpedic saturation phase (ip) (Coefma = 0). For a color version of the figure, see www.iste.co.uk/braudeau/hydro.zip*

The example shown in Figure 9.3 is taken from an Andosol sample from Martinique (no. 278, Plateau Perdrix, 30 cm), which is analyzed as part of the Martinique SIRS-Soils [BRA 07]. This is what explains the high water content in the reading range of the tensiometer and the elevated number of measurement points (292 points). The result is spectacular and the curves are practically superimposed with a CE h between the two curves at 2.5 hPa.

The results, put under the more explicit form of the theoretical curve's parameters: W_{misat}, W_{masat}, $\overline{E}_{mi}$ and $\overline{E}_{ma}$, are given in the "Results" table in Figure 9.3. These are the parameters for the theoretical curve drawn in the figure and adjusted to meet the closest (SCE minimum) on the measured curve. It has been shown that this theoretical curve remains valid well after the end of validity range inherent to the tensiometer, at least until 3,200 hPa, which is the suction value that corresponds to the air pressure of 10 bars in the porous plate press device of the pFs measurement [BRA 14b].

9.1.1.2. *2nd case: with the interpedic (ip) saturation phase: W_L << W_{Sat}*

The retention curve shows a sloping ledge or shouldering off more or less exactly at the suction $h_{ip}°$ (~100–200 hPa), as shown in Figure 9.4 for example. The non-zero value of $h_{ip}°$ that is read on the curve (at the beginning of the sloping ledge towards the decreasing water content) is then introduced into the adjustment conditions of the first table. As we have already seen, this value constitutes the lower limit of the adjustment range for the measured curve.

Then, we introduce the value for the upper limit of validity and the W'_{Sat} values that are read on the measured curve, just as we did in the previous case.

The seven parameters of the retention curve are determined in two stages; the first one using *Coefma* = 0 and the second one using *Coefma* = 1:

– 1st stage:

Coefma = 0: only the portion of the curve for which $h > h_{ip}°$ is adjusted ($h_{ip}°$ = 130 hPa in the example shown in Figure 9.4).

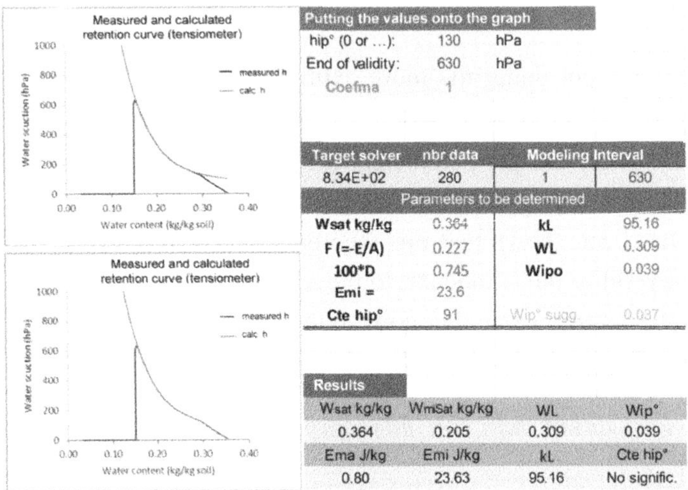

Figure 9.4. *Sheet for adjusting the retention curve. Graphic examples of the two adjustment stages (Coefma = 0: curve from the top and Coefma = 1: curve from the bottom); the tables display the values obtained after adjusting the 2nd stage. For a color version of the figure, see www.iste.co.uk/braudeau/hydro.zip*

Having input *Coefma* = 0, only the portion of the curve for which $h > \overset{\circ}{h}_{ip}$ is adjusted. In this example shown in Figure 9.4, $\overset{\circ}{h}_{ip}$*=130 hPa*. In this stage, the setup of the parameters is similar to the case described above of no-interpedic saturation phase, but with an extra parameter to determine $(Cte\ \overset{\circ}{h}_{ip})$ which is initialized by a value of $\overset{\circ}{h}_{ip}$. This constant actually replaces the parameters in the right hand side of the green table, namely k_L, W_L and W_{ipo} in modeling the WRC. These parameters will be determined in the next step, i.e. Coefma = 1.

The Excel solver is used to determine the four parameters of the left part of the green table: (F), $(100 \times D)$, $\bar{E}_{mi}$ and $Cte\ \overset{\circ}{h}_{ip}$. W'_{Sat} is not included in the adjustment and remains at its estimated value on the curve at *h=0 hPa*. The result of this adjustment using *Coefma* = 0 is shown in Figure 9.4 (top of the graph).

– 2nd stage:

Coefma = 1: putting *Coefma* to 1 has the effect of reactivating the parameters of the right side of the green table, and of canceling the role of Cte $h_{ip}°$ (as it becomes inactive in the table). In this way, the whole curve is used in the adjustment, from saturation at $h = 0$ until the end of validity range of the tensiometer readings.

Without changing the parameters of the first stage, we initialize the parameters for the interpedic phase, which are located on the right hand of the green table: $k_L = 50$; W_L is set to the water content read at the bulge; and W_{ip}^0 as the suggested value in the table.

The Excel solver is used for the second time, but with *Coefma* = 1, to determine the parameters of the interpedic phase k_L, W_L, and W_{ip}^0 also W'_{sat}. The group of seven parameters is shown in table 3, called "Results", in Figure 9.4. They are the results of adjusting the theoretical curve of $h = h_{mi} + h_{ip} = h_{ma} + h_{ip}$, (section 8.3.2) to the measured curve represented in the graph at the bottom of the figure.

Note that the passage through the first step (*Coefma* = 0) is useful to calculate the approximate value of W_{ip}^0, based on $\overset{\circ}{h}_{ip}$, which is suggested in the bottom right of the green table in the second stage of adjustment with *Coefma* = 1.

9.1.2. *Measured under air pressure on the porous plate press (pressure-based method), extending the WRC measurement beyond the 1,000 hPa tensiometric limit*

Pressure plate is widely used as a method for measuring the WRC at higher suctions by using a completely different principle (pressure-based method instead of suction-based method). Section 8.3.3 provides the information and equations needed to convert the measured data into retention pressure *h*. Despite being widely used, there are still some limiting factors once compared with the tensiometer readings, including the small number of readings, the lack of the precision in the WRC measurements near saturation and that the different points of soil suction on the measured WRC are taken from different soil samples (Figure 9.5). It is therefore impossible to confirm the presence of a possible interpedic saturation phase (ip) around the saturation even if the measurement points were taken in the area. That is why the parameter $Cte\ \overset{\circ}{h}_{ip} \geq 0$ becomes part of the variable parameters in addition to the four parameters of the water retention curve seen previously (with *Coefma* = 0): i.e. W_{Sat}, $F = -E/A$, $D = -\bar{E}_{ma}/A$ and $\bar{E}_{mi}$ in the green table. Therefore, the adjustment should focus on the points whose suction is greater than 150 hPa. The procedures for adjusting the modeled WRC to the measured data points of the pressure plate apparatus are shown below (see also Figure 9.5).

The first two columns in Figure 9.5 are filled in by copying and pasting the measurement data of the classic pF curve; the 1st column is the applied air pressure (in *hPa* or *mbar* or *cm* of water) and the 2nd column is the corresponding gravimetric water content $W\ (kg_{water}/kg_{solids})$. The saturated water content of the soil sample W'_{Sat} must be entered on the top of the 2nd column.

The following column, (*h* [hPa]), converts the applied air pressure on the sample in the porous plate press, in suction or water retention pressure, *h*, according to equation [8.18], and that depends on the temperature with which the measurement is carried out. This is 23°C by default but can be changed (cell K27 in the "Calculations" table of the Excel sheet).

Press. (hPa)	W (kg/kg)	h [hPa]	W_{ma} calc	W_{mi} calc	h_{mi} (hPa)	h_{ma} (hPa)	Carré Ec h
Saturation	0.448	2	0.22916321	0.219	0	0	
100	0.352	130	0.13725785	0.215	137	137	0.3
	0.344		0.12941504	0.214	158	158	
400	0.281	460	0.07404982	0.207	428	428	2.2
600	0.259	643	0.05737261	0.202	612	612	1.4
800	0.240	804	0.04371146	0.196	867	867	5.1
1000	0.233	948	0.03959609	0.193	979	979	1.0
1500	0.219	1253	0.03208255	0.187	1256	1256	0.0
2000	0.210	1502	0.02790326	0.182	1475	1475	0.5
4000	0.190	2200	0.02008462	0.170	2128	2128	2.4
10000	0.164	3278	0.01328767	0.151	3321	3321	0.6

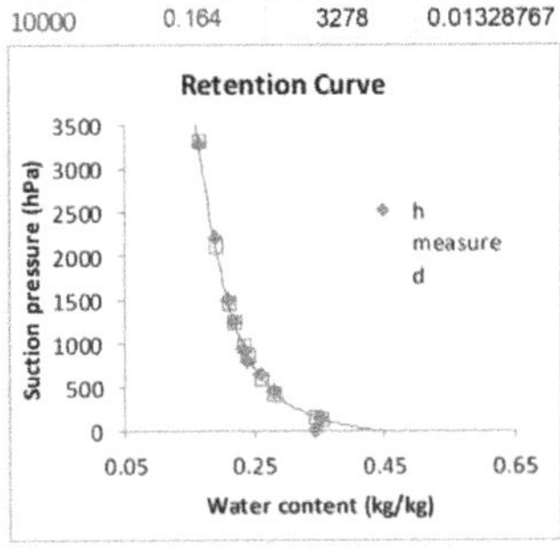

Ema/A =	-6.49E-03	A=	-7.22E+02
E/A =	-0.231	E/Ema =	36
Ema =	4.6849915	Wsat =	0.448
Wsat-F=	0.679	WM2=	0.219
delta =	0.24195348	WM1=	0.461
Temp (°C) =	23	Mw (kg/mole)	0.018
R (J/mole/K)	8.314	RT/M (J/kg)	136719

Figure 9.5. *Adjustment sheet for the retention curve measured with the porous plate-press (pFs). Explanations in the text*

After providing the sample name and potential observations, variable parameters of the green table (Figure 9.6) are determined as follows: *Wsat* is placed at its provided measured value; *F* is at the water content level at the beginning of the suction incline (between 200 and 400 hPa); 100 * *D* at 0.1 and E_{mi} at 100 and *Cte* h_{ip}° at 0.

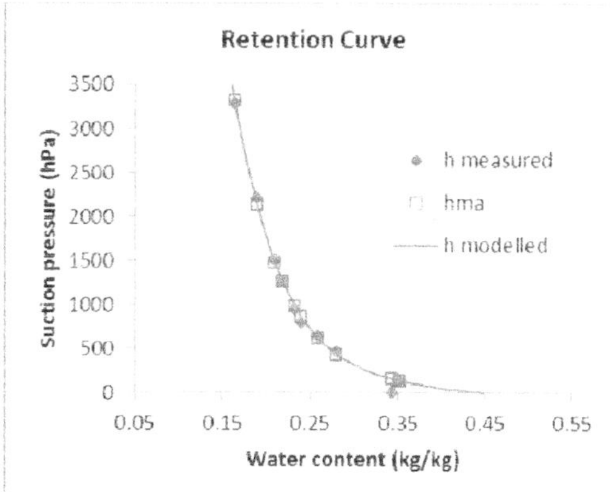

Variable Parameters				
Wsat (kg/kg)	F= (-E/A)	100*(-D)	Emi J/kg	cte hip°
0.448	0.23	0.64907483	162.3	0

Cell lim. n°	Cibles du Solver Targets	
3	SCE *h*	SCE *W*
12	13.5	0.00024031

Results				
Wsat kg/kg	WmiSat kg/kg	Ema J/kg	Emi J/kg	hip°
0.448	**0.219**	**4.6850**	**162.35**	**0.00**

Figure 9.6. *Graph, command tables and results of the adjustment sheet of the retention curve measured with the porous plate press*

We then introduce the adjustment range line numbers (in blue in the red table under "Cell lim. no" in Figure 9.6) and then the solver is activated to determine the parameters. We choose SCE *h* as the target (in the "Solver targets") in order to minimize, and just as with the parameter variables in the green table above, except for W_{sat} which stays fixed. If the solver does not converge (rare), we choose SCE *W* as a target and we finish in the second phase with SCE *h*.

It is possible to remove an outlier by removing the value of the corresponding pressure in column 1. In the example of Figure 9.6, the adjustment dealt with all of the measurement points, except the outlier at $h = 200\ hPa$ that has been deleted from the 1st column, "Press (hPa)" (Figure 9.5).

It is also possible to adjust the calibration range by changing the limits of the adjustment range as in the example of Figure 9.7, where the adjustment only dealt with the six points between 800 and 10,000 hPa (lines 7 to 12 selected in the middle table) with only the three parameter variables: F, D and E_{mi} (the parameters $Cte\ h_{ip}°$ and $Wsat$ are the same as in Figure 9.6).

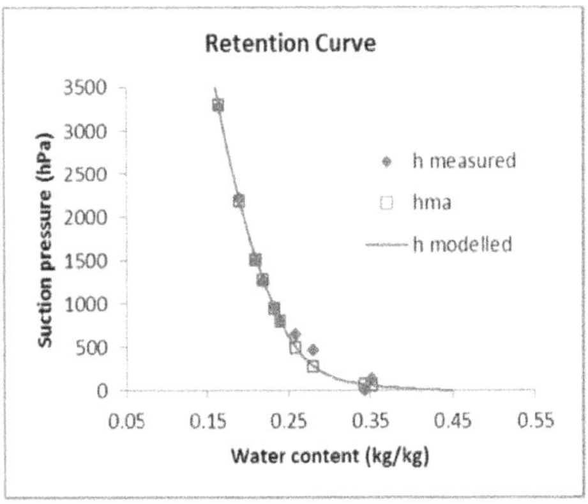

Variable Parameters				
Wsat (kg/kg)	F= (-E/A)	100*(-D)	Emi J/kg	cte hip°
0.448	0.27	0.20773129	138.2	0

Cell lim. n°	Cibles du Solver	
7	SCE *h*	SCE *W*
12	0.5	4.3269E-06

Results				
Wsat kg/kg	WmSat kg/kg	Ema J/kg	Emi J/kg	hip°
0.448	**0.261**	**1.0855**	**138.17**	**0.00**

Figure 9.7. *Same sample and analysis sheet as in the previous figure but only the six measurement points of 7 to 12 are adjusted (middle table)*

In each case, within the adjustment range, the difference between the measured and calculated water contents is at a maximum of $\pm$ 0.002 kg_{water}/kg_{solid}, which is a lot less than the resolution of the water content measurements from the lab sample.

9.2. The shrinkage curve

9.2.1. *Case of non-sigmoidal shrinkage curves*

When the aggregate structure is weak, the distinction between the micro-water on the inside of the primary aggregates and the macro-water in the inter-aggregate space surrounding the primary aggregates is no longer visible on the shrinkage curve. It can only be determined by measuring and modeling the retention curve. Therefore, we need the soil sample characteristic retention curve to be able to extract the parameters of its shrinkage curve. This method of characterization uses the recent thermodynamic equations originally established by Braudeau *et al.* [BRA 14a, BRA 14b] and associates the retention curve for the soil sample with its shrinkage curve. The WRC should be measured by the tensiometer or the porous plate press. These equations of WRC are recalled in section 8.3.4 and serve as a basis for programming the Excel sheets described below for determining the parameters.

The Excel sheet for adjusting the shrinkage curve is composed of two interconnected sheets, one of them, "CTENSIO", is for adjusting the retention curve, and the other one, "CRET", is for adjusting the shrinkage curve, using the parameters determined in the first.

The upper left side of the CRET sheet is shown in Figure 9.8. The first two columns show the measurement data values for W and V, then the calculation columns for the different variables entering the modeling: w_{re}, w_{bs}, W_{mi}, w_{st}, w_{ip}, etc. The column "*wst Exp*" is the water content (w_{st}) calculated according to the old equations of [BRA 04], to compare and obtain the parameter k_M of the old version. "*Vmod*" is the specific volume modeled and, "*Crmi*" is the shrinkage curve of the micro-porous phase, due to w_{bs}. These calculated variables according to W are found in both graphs present on the Excel sheet (Figure 9.8). In the first graph, we can distinguish the retention curve (in green, coming from the CTENSIO sheet), the modeled ("*Vmod*", in red) and measured (in dark blue) shrinkage curves, and the shrinkage curve is calculated on the clay-plasmic phase ("*Crmi*", the red dotted line). In the graph on the right, the different water types are represented according to the water content of the pedostructure, *W*. The two superimposed curves (columns "*wst Exp*" and *Wma*) show the resulting equivalence between the semi-empirical equation of $wst = Wma$ according to *W* used beforehand (parameters k_M, W_M) and the new equilibrium equations of *Wma* and *Wmi* (of which the parameters are *F* and *D*).

Three examples are presented below.

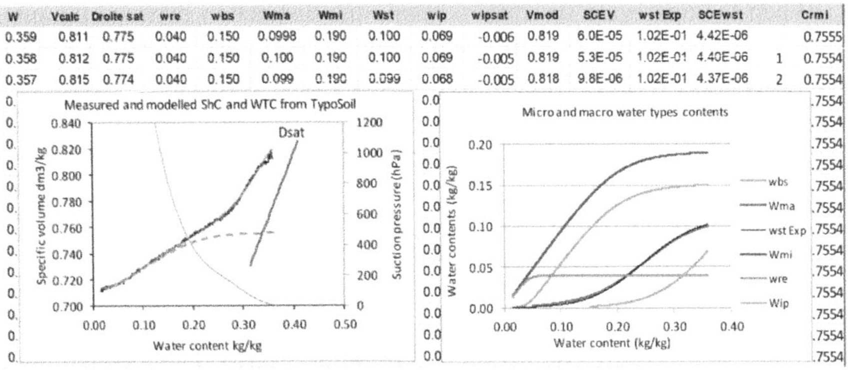

Figure 9.8. *CRET Sheet for adjusting the shrinkage curve based on the information given by the CTENSIO retention curve associated with it. For a color version of the figure, see www.iste.co.uk/braudeau/hydro.zip*

9.2.1.1. *The first example*

The soil sample named "Versailles 3C" is shown as example in Figure 9.9. This is a reconstituted soil cylinder made of a mixture of sieved aggregates: 72 % >200 µm and 28 % <50 µm packed in a cylinder of 5 cm height and 5 cm diameter.

9.2.1.1.1. Preliminary adjustment to the retention curve

The CTENSIO sheet is presented in the same way as the TENSIO sheet already shown (section 9.1.1, Figure 9.1). The data for *W* and *h* are input into the first two columns, either by copy–paste or in line with TypoSoil® result sheet, and the curve adjustment procedure is the same as that described previously in section 9.1.1. Figure 9.9 shows the adjustment result of the retention curve as well as the graph displaying the measured curve (black line) and the adjusted curve (red line).

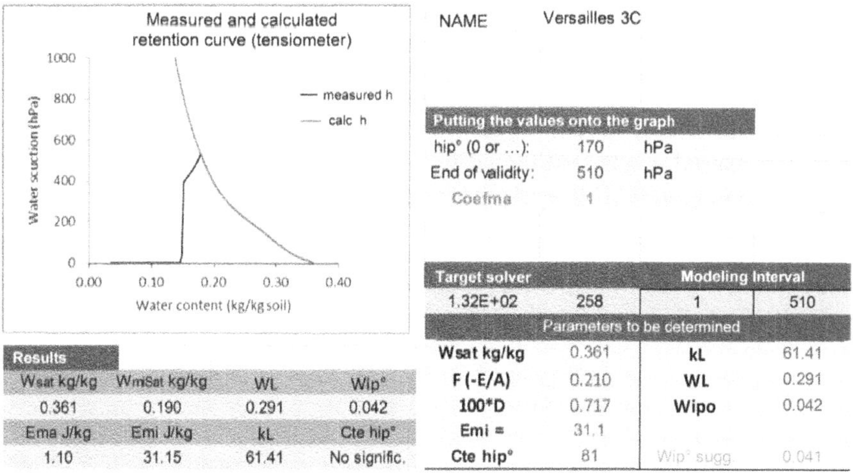

Figure 9.9. *Components of the CTENSIO sheet: adjusting the retention curve prior to the shrinkage cure. For a color version of the figure, see www.iste.co.uk/braudeau/hydro.zip*

9.2.1.1.2. Adjusting the shrinkage curve

After obtaining the water retention curve's parameters, we can use the CRET sheet for the adjustment of the shrinkage curve. The parameters coming from the retention curve (W'_{sat}, F, D and W_L) are displayed at the top of the green table (Figure 9.10). They will remain fixed while the shrinkage curve is being adjusted. Then we proceed by setting up the other parameters, finding out:

– the limits of the adjustment range: a lower limit (*W limit inf*) is chosen after the shrinkage limit, discarding all the excess data, as it is unnecessary for the adjustment. In the same way, an upper limit (micro W up *ShC*) is chosen in order to delineate the part of the curve to be adjusted, when we only want to take that which corresponds to the plasmic micro-porosity's area of influence (indicated by a dotted line) or also for eliminating the first measurement values, towards saturation, which are sometimes irregular. The Solver Target to choose will be between "*ShCmi*" (sum of the squared derivation between the limits W_{inf} and W_{sup}) and "ShC whole" (SCE between W_{sat} and W_{inf}) depending on the particular case;

– *Vo* and *ds*: the specific dry volume of the analyzed sample as read on the curve at the adjustment range limit: W_{inf} and the real density of the solid phase, measured or else estimated by the modeler at some % to the right of the saturation equation: $V = W/\rho_w + 1/ds$;

– the "Variable Parameters" table, by putting kN/100 to 1, W_{miN} at the value point W_N read on the curve, K_{bs} at 0.5, *kL/100* at the value suggested by CTENSIO (Figure 9.9), K_{st} at 0 and K_{ip} at 1.

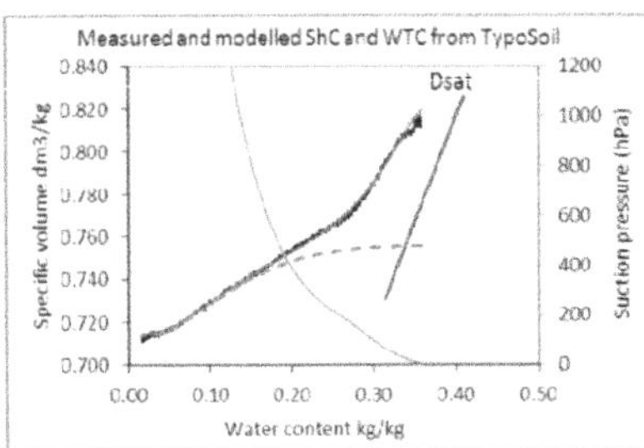

Name	Versailles 3C
W limit inf	0.033
W up ShCmi	0.338
coefma	1

Observed on the curve	
Vo	ds
0.714	2.4

Provided by the retention curve			
Wsat	F (-E/A)	100*D	WL
0.361	0.210	0.717	0.300

Results				
coefma	Wsat	F=(-E/A)	D (=-Ema/A)	
1.00	0.361	0.210	0.007167	kgw/kgs
V0 dm3/kgs	Kbs'	Kst	Kip	
0.714	0.276	0.000	1.000	dm3/kgw
Vs	kN	kM	kl	
0.42	124.6	-25.2	22.8	kgs/kgw
	WmiN	WmiSat	WL	
	0.040	0.191	0.300	kgw/kgs

Solver Targets			
ShC micro	Whole ShC	SCE kM	kM
0.00047453	0.00089614	0.00124	25.2

Variable Parameters				
ShCmi parameters		ShC Wip parameters		Val suggérée
kN/100	1.25	kL/100	0.23	0.61
WmiN	0.040	Kst	0.00	
Kbs'	0.276	Kip	1.00	

Figure 9.10. *Tables of command and results of the CRET sheet as well as graphs of the measured and modeled shrinkage and retention curves, after having adjusted them both. For a color version of the figure, see www.iste.co.uk/braudeau/hydro.zip*

The solver is then activated to determine the variable parameters of the green table (at the bottom). In the present case, it is done in one run, by only taking *kN/100*, K_{bs}*'*, *WL* and *kL/100* as the variable parameters, K_{st} and K_{ip} having been fixed to 0 and 1. W_{minN} has also been fixed to its value read on the curve at point N as the range is very short and does not allow variation. The values for W_L and k_L given by adjusting each of the two curves, can be slightly different, but are still comparable.

9.2.1.2. *The second example*

The second example (Figure 9.11) is that of a reconstituted soil sample (sieved at 2 mm) from Qatar, taken from Assi *et al.* [ASS 14].

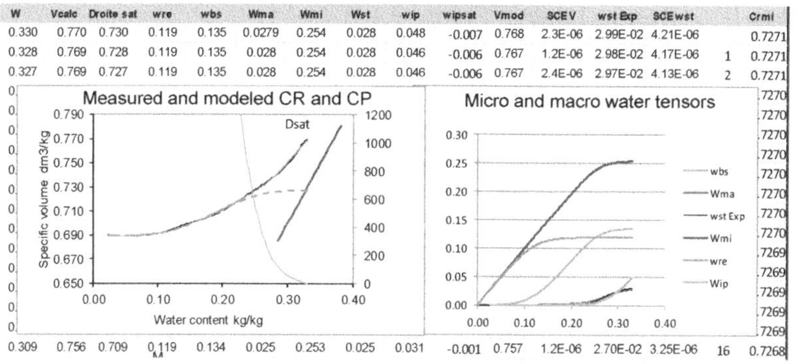

Figure 9.11. *CRET sheet. Example of adjustment of two associated shrinkage and retention curves; case of Qatar Rodah soil. For a color version of the figure, see www.iste.co.uk/braudeau/hydro.zip*

9.2.1.2.1. Preliminary adjustment

No particular point is clearly visible on either of the two curves (shrinkage or retention) in Figure 9.11. However, on the shrinkage curve, we can guess point N to be at about 0.10 and point L to be at about $W = 0.30$ kg/kg.

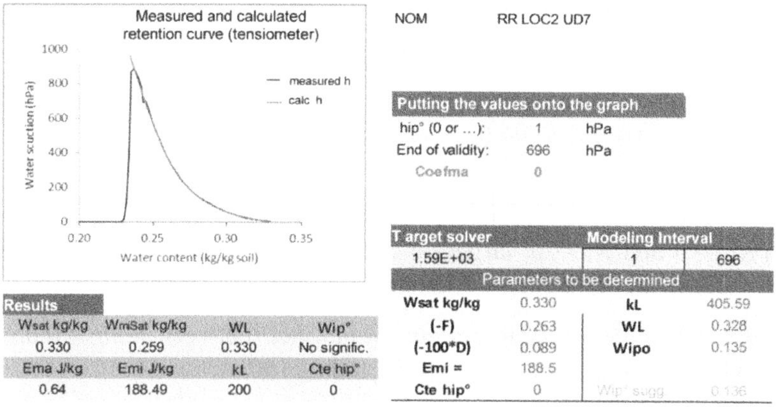

Figure 9.12. *CTENSIO Excel sheet: Adjusting the associated retention curve; case of Rodah Qatar soil. For a color version of the figure, see www.iste.co.uk/braudeau/hydro.zip*

As the retention curve does not display an interpedic (ip) saturation phase, it is processed using *Coefma* set at 0. It therefore contains only four parameters, W_{sat}, F, D and E_{mi}, of which the first three will be used for the shrinkage curve. The value of W_L transmitted onto the CRET sheet will be equal to W_{sat} (Results table in Figure 9.12).

9.2.1.2.2. Adjusting the shrinkage curve

The distinction of point L on the outskirts of $W = 0.30$ kg/kg (close to W'_{sat}) (Figure 9.13) induces us to think that it is the interpedal water (W_{ip}), which causes the shrinkage of the soil sample (with $K_{ip} = 1$) while being in competition with W_{ma} during evaporation whose departure does not cause shrinkage ($K_{st} = 0$).

We begin by finding out the adjustment range limits (for the micro and whole ShC targets) of the corresponding cases, and then we introduce the values for *Vo* and *ds* as we did before.

Because we suppose that there must be an important action for W_{ip} on the shrinkage curve, although one is not visible on the retention curve, we change the value for W_L provided by this last one in the upper green table (which was equal to W_{sat}). By putting W_L at the estimated value read on the curve (0.285), *Coefma* is automatically put to 1 (which in the calculations, activates the role of W_{ip} and in particular, the appearance of parameters W_L and W'_{sat} as much as $W_{ipSat} = (W'_{sat} - W_L)$ and $W_L = (W_{maSat} + W_{miSat})$, refering to equation [8.7] to [8.13].

The variable parameters from the bottom green table are set up as before: W_{miN} is placed at the value of W_N read on the curve, *kN/100* at 1, *kL/100* at 1, K_{bs}' at 0.5, K_{st} at 0 and K_{ip} at 0.

The Excel solver is used to determine the parameters in two steps:

1) $K_{ip} = 0$ and $K_{st} = 0$: target on *ShC micro* (between $Winf = 0.033$ and $Wsup = 0.2$) with the variable parameters: $k_N/100$, W_{miN} and K_{bs}';

2) $K_{ip} = 1$ and $K_{st} = 0$: target on *Whole ShC* (between $Winf = 0.033$ and $Wsat$) with the variable parameters: W_L and $k_L/100$.

9.2.1.2.3. Results

The group of results from the two pages (CTENSIO and CRET) is displayed in the "Results" table of Figure 9.13. In the table, we note that k_M and $W_{miSat} = W_M$, are present, parameters from the old equations for modeling the shrinkage curve in the M part of the curve. These two parameters have been replaced by the two new parameters D and F from the micro-and macro-water types at equilibrium. We have seen how W_{miSat} can be calculated based on F and D and W_L or W_{sat} (equation [8.16]). As for k_M, we can obtain it by adjusting the curve $W_{ma}(W)$ (new equation [8.4a]) by the equation (older) of w_{st} analogous to equation [8.10]; which is done by the solver having the target "*SCE kM*" and "k_M" as a variable parameter just next to it.

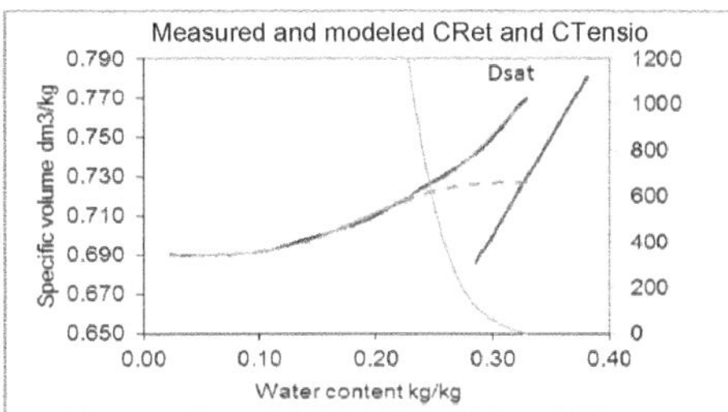

Name RR LOC2 UD7

W limit inf	0.033	coefma
W up ShCmi	0.2	1

Observed on the curve	
Vo	ds
0.689	2.5

Provided by the retention curve			
Wsat	F (-E/A)	100*D	WL
0.330	0.263	0.089	0.285

Results				
coefma	Wsat	F=(-E/A)	D (=-Ema/A)	
1	0.330	0.263	0.000891	kgw/kgs
V0 dm3/kgs	Kbs'	Kst	Kip	
0.689	0.282	0	1	dm3/kgw
Vs	kN	kM	kl	
0.40	42.7	-62.8	44.9	kgs/kgw
	WmiN	WmiSat	WL	
	0.119	0.255	0.285	kgw/kgs

Solver Targets			
ShCmicro	Whole ShC	SCE kM	kM
6.9475E-05	0.00014929	0.00039	62.8
Variable Parameters			
ShCmi parameters		ShC Wip parameters	
kN/100	0.43	kL/100	0.45
WmiN	0.119	Kst	0.00
Kbs'	0.282	Kip	1.00

Figure 9.13. *CRET Excel sheet. Tables of command and results as well as graphs of the measured and modeled shrinkage and retention curves after adjustement of the shrinkage curve; case of Qatar Rodah soil. For a color version of the figure, see www.iste.co.uk/braudeau/hydro.zip*

According to the "Results" table, the number of characteristics for the shrinkage curve of the pedostructure is 12 at the most: Vo, V_s, W'_{sat}, F, D, K_{bs}', K_{st}, K_{ip}, k_N, W_{miN}, k_L and W_L. The number is actually 10 if we consider that K_{st} and K_{ip} have the values of 0 and 1 in practically every case. The number is further reduced to 8 if we consider that V_s and W'_{sat} are not part of the parameters in the shrinkage curve equations.

9.2.1.3. *The third example*

The soil sample is the same as that used in the 1st example (Versailles soil sifted at 2 mm), but the aggregate mixture this time is: 54% > 200 μm and 46% < 200 μm.

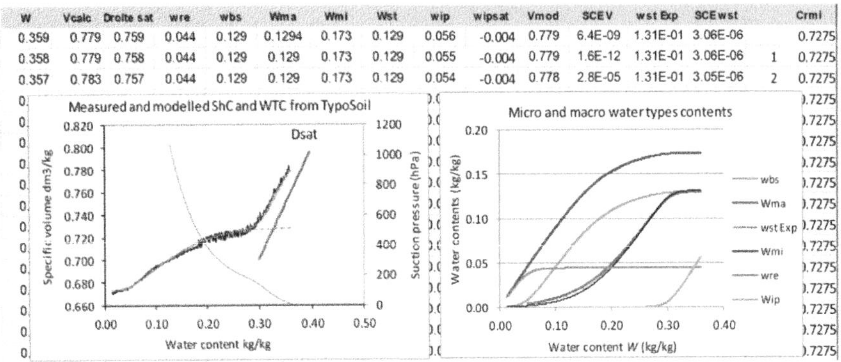

Figure 9.14. *CRET Excel sheet for adjustement of the shrinkage curve; case of "Versailles soil 4D". For a color version of the figure, see www.iste.co.uk/braudeau/hydro.zip*

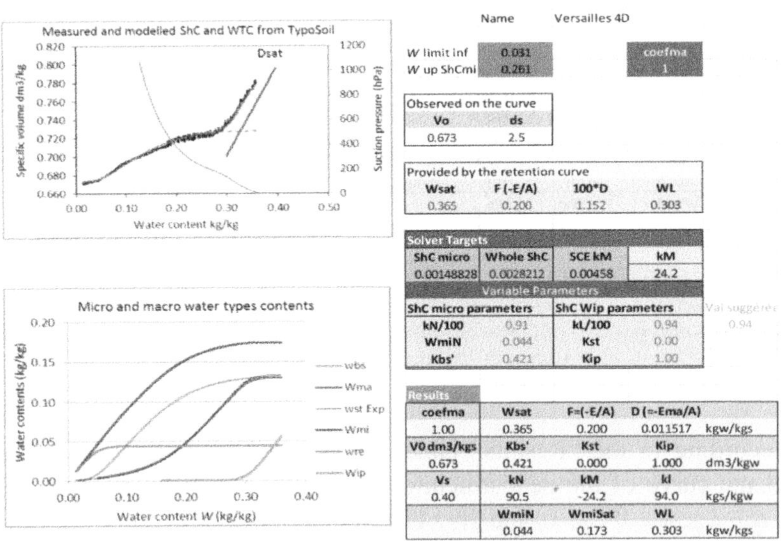

Figure 9.15. *CRET Excel sheet. Tables and graphs after adjustement of the shrinkage curve for the sample "Versailles 4D". For a color version of the figure, see www.iste.co.uk/braudeau/hydro.zip*

Once again, an associated CTENSIO sheet is used to adjust the retention curve prior the shrinkage curve. The tensiometric curve shows an area of interpedal (ip) saturation (Figure 9.14), which will therefore be processed like in the "2nd case" shown in section 9.1.1. First of all, we will use the *Coefma* = 0 mode (variables F, D, E_{mi} and *Cte* $h_{ip}°$) and then *Coefma* = 1 mode, after setting $W_{ip}°$ to the suggested value in the margin, using the other parameters to be determined: W_L, k_L, $W_{ip}°$ and W'_{sat}. The result is excellent and the values found for parameters F, D, W_L, k_L, $W_{ip}°$ and W'_{sat} are transmitted onto the CRET sheet in order to adjust the shrinkage curve (Figure 9.15).

This can be done in two steps, by turning the solver: 1) on the target: *ShC micro* (setting K_{ip} = 0 and the lower and the upper limits as 0.031 and 0.261, respectively), and with the variables k_N/100, W_{minN} and K_{bs}'; 2) on the *Whole ShC* (setting K_{ip} = 1,) and with the variables W_L and k_L/100 (which, once obtained, we will be able to compare with the values given by the tensiometric curve).

The two curves fit almost perfectly (Figure 9.15, the parameters found on the two curves W_L and k_L are very close) and the shrinkage curve is simulated in every part. The water distribution according to the total water content W is shown in the graph at the bottom of Figure 9.15. The curves are calculated according to the parameters given in the table called "Results", located on the side.

As with the previous examples, here again we note that the shrinkage is only due to W_{ip} and W_{mi} (because K_{st} = 0). In fact, W_{ma}, which compensates for W_{mi} with which this is in tension equilibrium in the pedostructure, does not generally provoke shrinkage (or swelling). The two water types that provoke the shrinkage/swelling are w_{bs} in the micro-porosity and W_{ip} in the macro-porosity. These examples confirm what was earlier said in theory, namely that in general, the shrinkage curve equation is:

$$\bar{V} = \bar{V}_0 + K'_{bs}(W_{mi} - w_{re}) + W_{ip}, \qquad [9.1]$$

for which the parameters are: $\bar{V}_0$, K'_{bs}, k_N, W_{miN}, D, F, k_L and W_L; the last four being in common with the retention curve.

9.2.2. *Case of the sigmoidal shrinkage curves*

9.2.2.1. *Presenting the Excel sheet for modeling CR alone*

As we have already mentioned in the introduction (Chapter 7), it is only the sigmoidal-type shrinkage curves which have a well-developed aggregate structure that can be processed according to the micro/macro pedostructure model of Braudeau *et al.* [BRA 04], without the need to use the retention curve to determine the parameters for the equilibrium equation of W_{mi} and W_{ma} (k_M and W_M or F and D).

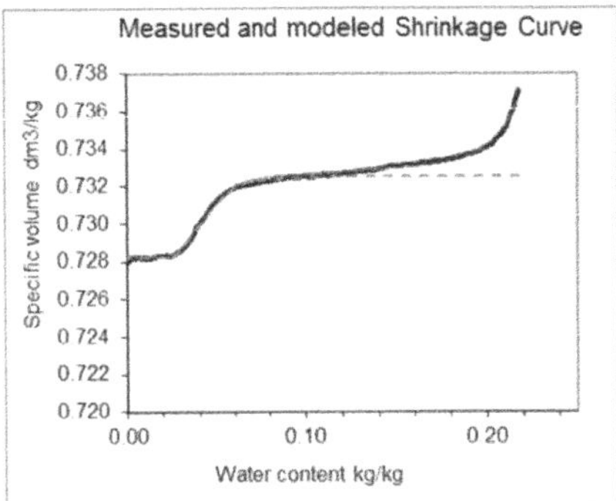

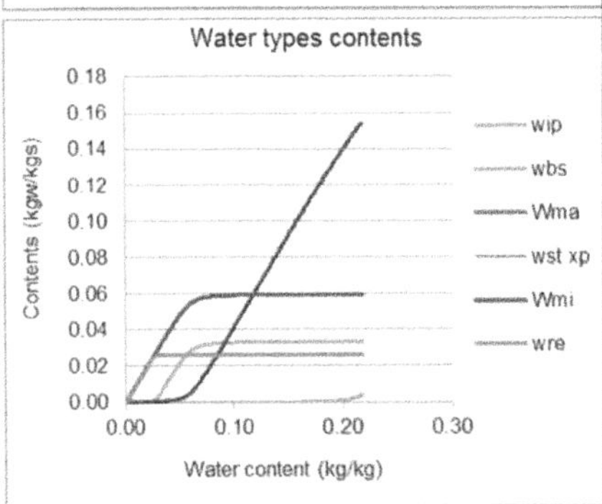

	Sample	NP5H20	
Range	**ShCmi**	**Kbs slope**	**Kbs obs**
W limit inf	0.006	0.040	**0.130**
W limit sup	0.116	0.045	
Vo	**ds**	**Wsat**	**WF**
0.728	2.4	0.240	0.3
Variable Parameters			
ShCmi parameters		**ShC Wip parameters**	
kN/100	7.08	**kL/100**	0.83
WmiN	0.026	**WL**	0.23
F (-E/A)	0.060	**Kst**	0.005
100*D	0.051	**Kip**	1.00
Solver Targets			
ShCmi	**Entire ShC**	**SCE kM**	**(-kM)**
3.34E-06	5.9067E-06	0.00002	122.0
Results	**coefma**	**F=(-E/A)**	**D (=-Ema/A)**
	1.00	0.060	0.000506
V0 dm3/kgs	**Kbs** dm3/kgw	**Kst** dm3/kgw	**Kip** dm3/kgw
0.728	0.130	0.005	1.000
Vs dm3/kgs	**kN** kgs/kgw	**kM** kgs/kgw	**kL** kgs/kgw
0.4167	708.1	-122.0	82.9
Wsat kgw/kgs	**WmiN** kgw/kgs	**WmiSat** kg/kg	**WL** kgw/kgs
0.240	0.026	0.060	0.229

Figure 9.16. *CRET Excel sheet for the sigmoidal shrinkage curves adjustement; case of a ferruginous soil from Senegall. For a color version of the figure, see www.iste.co.uk/braudeau/hydro.zip*

Therefore, the Excel sheet dedicated to processing the sigmoid shrinkage curves has no associated TENSIO sheet. It is a CRET sheet, which contains the same elements (variables columns, tables and curves) as that used in the previous case (for example, Figures 9.14 and 9.15). However, the tables for setting up the parameters (the three first ones starting from the top, Figure 9.16) are shown differently because of two reasons. Firstly, K_{bs}' is no longer a parameter to be determined, but instead it is to be fixed at the gradient value of K_{bs} observed on the curve. And secondly, because the parameters D and F are part of the variable parameters to be determined by the Excel solver. Setting up the Excel sheet before using the solver is done like this:

First of all, we fix the curve adjustment ranges as previously defined (see the first example of the previous section 9.2.1), by finding out the cases "*W limit inf*" and "*W limit sup*" on the red table, under the "*ShCmi*" *umbrella*. At the same time as we choose "*W limit inf*" on the curve, we also read the specific volume *Vo* which we introduce into the parameters table. Cases *ds*, W_{sat}' and W_F from the table are found just in the same way as the previous examples. Gradient K_{bs} from the "basic" shrinkage phase (between points B and C, Figure 8.1) is automatically calculated by placing "K_{bs} *slope*" under the water content "*W limit inf*" and "*W limit sup*", as read on the curve at the two points B and C, which delineate the basic shrinkage phase (bs) (Figure 8.1).

The variables to be determined by the solver are then set up in the green table in the same way as the previous examples. If there is no (ip) phase, K_{ip} is fixed at 0 and W_L is fixed at W_{sat}' (which is generally taken as the first measurement datum). The water content at point F, W_F, is estimated at the meeting point of the curve with the saturation line; it must always be larger than W_{sat}' by some %. If the shrinkage curve has an (ip) phase, the adjustment needs to be done in two steps. A first adjustment is made with Solver Target of "*ShCmi*", and for variable parameters, those from the left column (*kN/100*, W_{miN}, *F* and *100 * D*), K_{ip} having been previously set to 0. The second adjustment affects the whole curve (target: *Entire ShC*) with K_{ip} fixed to 1 and the variable parameters being: *F*, *D*, *kL/100* and *WL*; K_{st} is generally equal to 0 but can potentially take a value, such as in the present case.

We end up with a curve adjustment such as that shown in Figure 9.16, but we can go further by giving the results of interpreting the shrinkage curve according to the SSM (structural shrinkage model) established by Braudeau [BRA 88a, BRA 88b] and Braudeau and Bruand [BRA 93] for the sigmoid shrinkage curves.

9.2.2.2. *Interpreting the curve according to the SSM*

We have seen that in the method for processing the sigmoidal curves, we identified K_{bs}' with the slope K_{bs} that we needed to calculate directly from the curve in order to fix it as a parameter. This goes back to considering that, in the evaporation cycle, all macro-water (W_{ma}) disappears from the inter-aggregate porosity starting from point C (Figure 8.1), while the primary aggregates remain saturated until point B, the entry point of air into their plasmic micro-porosity. Thus, we join up the basic hypothesis of the SSM. This is practically the case for the samples with a well-developed fine aggregate structure (e.g. in kaolinite, halloysite and illite soils), which show a clear sigmoid curve where the shrinkage phases due to the water from the two porous spaces: intra- (micro) and inter- (macro) primary aggregates, are well distinguished. Various different works have used the interpretative capacity of this model [BRA 99, BRA 04, BRA 05, BOI 04, BOI 06, COL 96]. Therefore, the Excel sheet dedicated to the shrinkage curve provides at the same time as the adjustment results: 1) the parameters for the shrinkage curve (Figure 9.16); 2) the water content at the transition points of the shrinkage phases defined in the structural shrinkage model l (points A, B, C etc., Figure 8.1) as well as 3) the micro- and macro-water and air volumes of the pedostructure at these particular points (Figure 9.17).

	W	Vmod	Vsolid/V	Wre/V	Wbs/V	Vpma/V	Wma/V	Wip/V	Air/V
WA	*0.017*	0.728	0.705	0.0233	0.0000	0.271	0.0003	0.000	0.27
WN	*0.026*	0.728	0.705	0.0351	0.0018	0.258	0.0006	0.000	0.26
WB	*0.040*	0.730	0.704	0.0361	0.0185	0.242	0.0014	0.000	0.24
WC	*0.045*	0.731	0.703	0.0360	0.0253	0.236	0.0021	0.000	0.23
WM	0.060	0.732	0.702	0.0360	0.0390	0.223	0.0077	0.000	0.22
WD	*0.070*	0.732	0.701	0.0359	0.0428	0.220	0.0188	0.000	0.20
WE	*0.189*	0.734	0.700	0.0359	0.0453	0.219	0.1779	0.001	0.04
WL	0.229	0.746	0.688	0.0353	0.0446	0.232	0.2206	0.017	-
Wsat	0.240	0.754	0.682	0.0349	0.0442	0.239	0.2218	0.027	-
Vsolid	*0.514*								

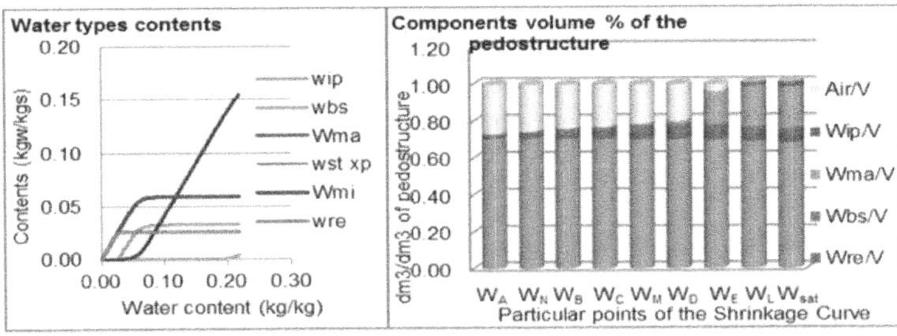

Figure 9.17. *Example of results display from the sigmoidal shrinkage curve analysis, according to the Structural Shrinkage Model (SSM). For a color version of the figure, see www.iste.co.uk/braudeau/hydro.zip*

The significance of these points and relations, which allow us to position the particular points of the curve according to the SSM, are given in Table 9.1, according to the articles written by Braudeau and Bruand [BRA 93] and Braudeau *et al.* [BRA 04].

Variable	Meaning	Calculation
W_N	Equivalent to the minimum micro-porous volume $Vp_{mi°}$; this is also the water content W_{re} at saturation: w_{reSat}.	W_N, W_M, W_L, as well as k_N, k_M and k_L, are part of the group of parameters that are determined by adjustment. The values of the corresponding specific volumes V_N, V_M and V_L are read on the curve at the corresponding water contents.
W_M	Equivalent to the micro-porous volume at saturation $Vp_{miSat} = W_{miSat}/\rho_w$. ($W_M - W_N$) represents the maximum swelling capacity of the clay plasma: w_{bsSat}. $Vp_{miSat} - Vp_{mi°} = w_{bsSat}$.	
W_L	Equivalent to the total porous volume of the pedostructure diminished by the interpedal saturation water W_{ip}. $W_L - W_M = W_{maSat} = Vp_{maSat} - W_{ipSat}$.	
W_{sat}'	Total water content at saturation: $W_{maSat} + W_{ipSat}$.	Estimated, must be equal or higher to the first value of measured W and to W_L.
W_A	Water content at the shrinkage limit.	$W_A = W_B - (V_B - V_o) * 1.718/K_{bs}$.
W_B	Water content at the micro air entry point.	$W_B = W_N + 3.46\ \mathrm{Ln}(2)/k_N$ or = "*W lim inf*" if W_B higher.
W_C	Water content at the beginning of the normal shrinkage phase of slope K_{bs}.	$W_C = W_M + 3.46\ \mathrm{Ln}(2)/k_M$ or = "*W lim sup*" if W_C lower.
W_D	Water content at the beginning of the effective shrinkage of the primary aggregates (at point D of the curve).	$V_D = Vo + K_{bs}(W_M - W_N)$. $W_D = W_C + 1.718(V_D - V_C)/K_{bs}$.
W_E	Water content at the point (E), lower limit of the interpedic shrinkage phase W_{ip}.	$W_E = W_L - 4.8\ \mathrm{Ln}(2)/k_L$.

Table 9.1. *Meaning of parameters and state variables at transition points of the shrinkage phase, and means of obtaining them*

Figure 9.18 shows examples of shrinkage curves of four different soil types, from which we can deduce the volume distribution of the different components of the pedostructure organization (solid phase of the structure, the three water types: W_{mi}, W_{ma} and W_{ip} and air) at the different particular points of the shrinkage curve, according to the SSM. Here, we understand more easily what makes the difference between one soil type and another in terms of water and air volume available to the plant and fauna of the soil.

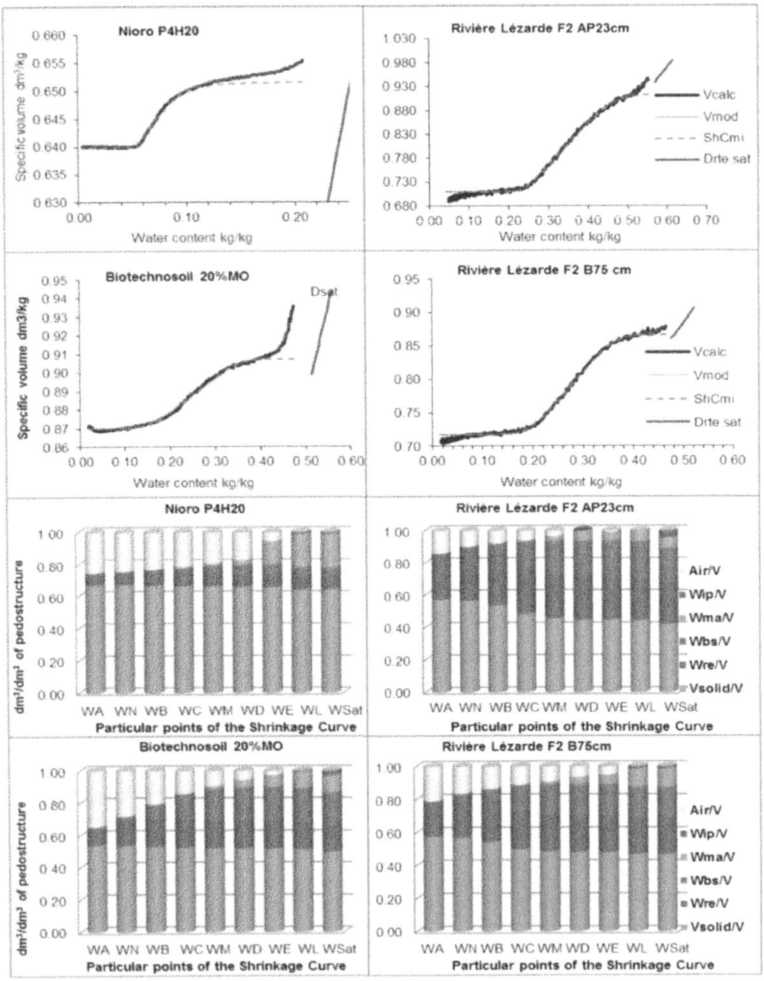

Figure 9.18. *Measured and modeled shrinkage curves of the four soil samples (top graph) and their interpretation is given in terms of volume percentage of the different phases: solid, liquid and air, which make up the pedostructure, calculated according to the SSM (bottom graph). For a color version of the figure, see www.iste.co.uk/braudeau/hydro.zip*

These soil samples were obtained from various studies conducted in Senegal (Nioro) [BRA 05], Martinique (Lézarde Riviera) [BRA 07] and France (Biotechnosol) [DEE 15].

9.3. The hydric conductivity curve of the pedostructure

9.3.1. *Description of the Excel sheets*

According to what was shown in section 8.3.5, determining the hydric conductivity k_{ps} by using the envisaged device (Figures 8.5 and 8.6) is only possible if we determine the parameters of the sample curve ($h(W)$), in a way in which the reading of the two tensiometers precisely informs us of the quantities W_{ma}, W_{mi} and the sum of the two, W, at the level of each of them.

When the hydric conductivity of the Martinique soils [BRA 07] was measured, the hydrostructural characterization of the measured hydric conductivity was not exercised.

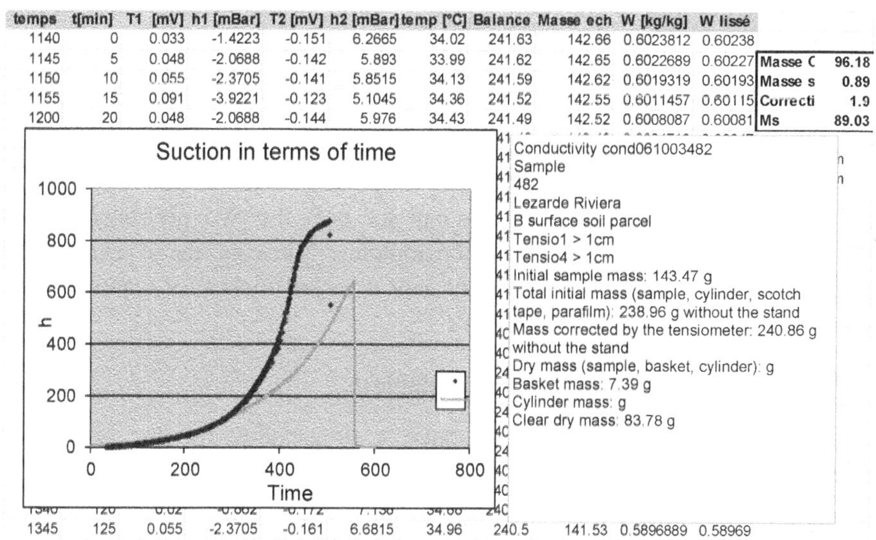

Figure 9.19. *Sheet of hydric conductivity measurements, according to the device described in section 8.3.5. For a color version of the figure, see www.iste.co.uk/braudeau/hydro.zip*

In order to overcome this, we have assumed (hypothesis 1, section 8.3.5) that the water content (W_2) at the level of the lower tensiometer T2 was approximately equal to the sample's total water content (W_{tot}), although we know that there is a vertical gradient of water content in the sample due to gravity, and where the water begins on the upper surface, due to evaporation. This is why, adjoined to the hydric conductivity calculation sheet, there is a sheet for adjusting the tensiometric measurements, like those presented above and called "*Tensio2 inf*". This allows us to obtain the modeling parameters of *h2* according to the total *W* content. These parameters are considered to be the hydrostructural characteristics of the sample, as we assume that the curve $h_2(W_{tot})$ is a good approximation of $h(W_{tot})$, the characteristic retention curve when measured with Typosoil®, for example, where evaporation occurs on all sides of the sample at once. Once the parameters have been obtained with the "*Tensio2 inf*" sheet, they are automatically input into the green table called "Parameters of the retention curve" on the calculation sheet, Figure 9.20. These are then used to calculate the water content W_{ma1} and W_{mi1} corresponding to h_1 on the T1 tensiometer (equation [8.3]). A second Excel sheet for adjusting the retention curve, called "*Tensio1 sup*", has been added in order to adjust the curve $h_1(W)$ provided by the upper tensiometer. This is not for calculating the curve's parameters, but to smooth the curve $h_1(W_{tot})$ and to be able to use the simulated curve which is continuous and not limited to 800 hPa. This is instead of the measured one which always displays micro-variations which disturb the calculations and the validity range ends at about 600–800 hPa. In the graph shown in Figure 9.20, we can see how the two simulated curves extend the validity range of the tensiometer, allowing us to extend the calculation of the hydric conductivity over a larger interval of water content than that on the tensiometric readings.

We therefore have the following variables:

– Measured: every 5 min, we have a time value for h_1, h_2 and W (Figure 9.19).

– Smoothed: from these data columns, we obtain the measured "smoothed out" variable columns: *h1mod* and h2mod by the means of their theoretical curve and *Wtotal* by its mobile average (Figure 9.20).

– Calculated: W_{ma2}, W_{ma1}, W_{mi2}, W_{mi1}, $-dW_{ma1}/dt$, $-dW_{ma2}/dt$, $-d_{W1}/dt$, $-dW_{mi1}/dt$, $-dW_{mi2}/dt$, $-d_{W2}/dt$, $-d_{W12}/dt$, W_{12}, W_{ma12}, Curve g, Curve2, Ln(Curve g).

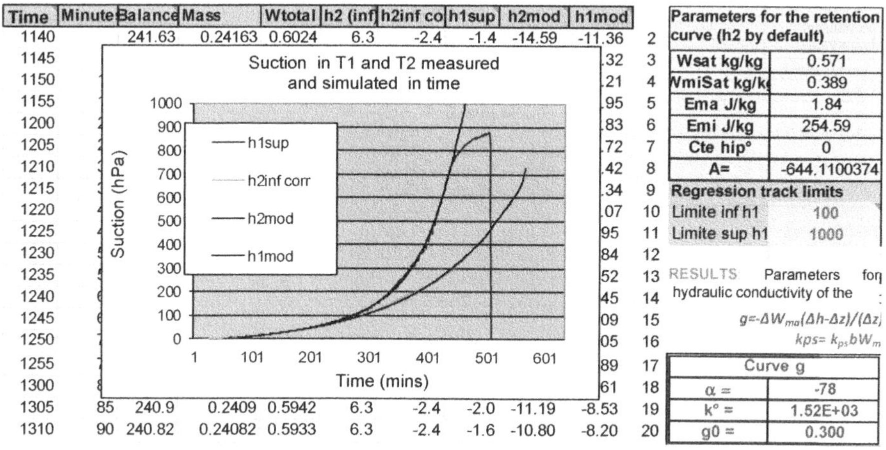

Figure 9.20. *Left side of the sheet for calculating the hydric conductivity, including the smoothed measured variables, the graph of the measured and the modeled curves h_1 and h_2, and the table (in green) of the characteristic parameters of the sample coming from curve $h_2(W_{tot})$ processed on the "Tensio2 inf". For a color version of the figure, see www.iste.co.uk/braudeau/hydro.zip*

In this way, the Excel spreadsheet for determining the hydric conductivity of the pedostructure is composed of five sheets:

1) the measurement data sheet (Figure 9.19) displaying the measured variables;

2) the calculation sheet that groups together: (1) the smoothed measurement data (Figures 9.20) and, based on these, (2) the calculated variables cited above and 3) the curves and the regression calculation, which provide the researched conductivity parameters $k_{ps}(W_{ma})$;

3) the sheet for adjusting the data of the lower tensiometer "*Tensio2 inf*" which provides the hydrostructural parameters of the shrinkage curve characteristic to the sample;

4) the sheet for adjusting the data of the upper tensiometer "*Tensio1 sup*" in order to "smooth" the curve $h1(Wtot)$.

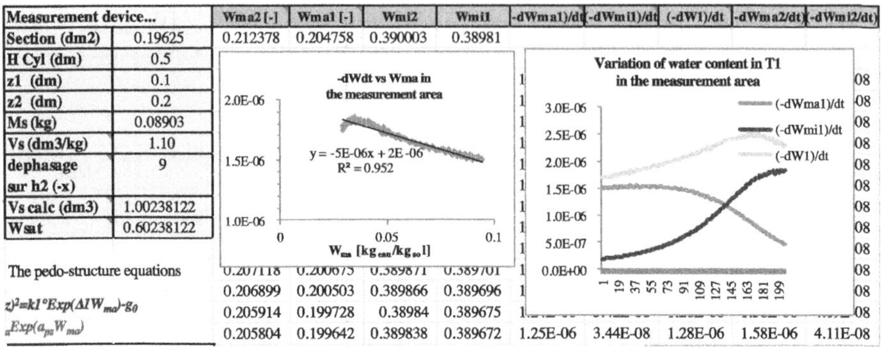

Figure 9.21. *Right side of the calculation sheet including the table of information on the measurement device, the columns of calculated variables necessary for calculating the hydric conductivity and some test curves. For a color version of the figure, see www.iste.co.uk/braudeau/hydro.zip*

9.3.2. *Procedure*

After introducing the measurement data into sheet (1), we must check, on the calculation sheet (2), the information coming from sheet (1) that is inscribed into the table "Measurement device". We then carry out the retention curves (*h2* and *h1*) adjustments successively (Figure 9.22) by using sheets (3) and (4): "*Tensio2 inf*" and "*Tensio1 sup*", respectively, as described above (section 9.1.1).

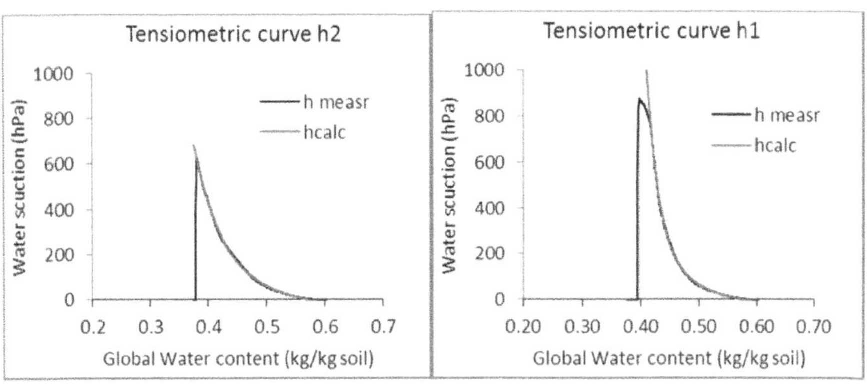

Figure 9.22. *Adjusting the tensiometric curves h1 and h2 according to the global water content of the sample. The deducted parameters of the h2 curve are considered to be the hydrostructural parameters of the sample. For a color version of the figure, see www.iste.co.uk/braudeau/hydro.zip*

The characteristic parameters of the $h_2(W)$ curve are automatically registered in the green table, called "Parameters of the retention curve" on the calculation sheet (2) (Figures 9.20 and 9.23), and the columns of the different variables depending on the parameters are instantly updated.

By observing the modeled curves of the two tensiometers shown in the graph of Figure 9.20, we can choose the regression range limits that we introduced into the corresponding table (below the parameters table). These limits define the validity range for the calculation of the two terms [8.30] of the Richards equation [8.27]. The lower limit is defined by a difference ($h_1 - h_2$) which must be sufficient (higher than 20 hPa) and the upper limit must be chosen on *h1mod*, a little after the capillary rupture of the tensiometer. In the example given here (Figures 9.20 and 9.23), the chosen regression range limits are from 100 to 1000 hPa. The two linear regressions between Ln($-g$) and W_{ma}, on the one side, and dW/dt and W_{ma}, on the other side, are automatically shown in the two tables at the bottom of Figure 9.23, in the desired validity range (cells 273–463).

9.3.3. *Results*

We recall that we have the following relationships (section 8.3.5):

$$\frac{dW}{dt} = -\rho_w \bar{V} \left(\frac{d(k_{ps} g)}{dW_{ma}} \right),$$

where the function $g(W_{ma})$ has the following expression (equation 8.32):

$$g = \left(\frac{\Delta h - \Delta z}{\Delta z} \right) \frac{\Delta W_{ma}}{\Delta z} = -k_1 Exp(\alpha_1 W_{ma}) + g_0$$

where $dW/dt = (aW_{ma} + b)$.

The parameters of $g(W_{ma})$ and $k_{ps}(W_{ma})$ are given in the yellow table in the middle of Figure 9.23. The constant g_0 is weak in front of the exponential and becomes imprtant only near the sample's saturated state, outside of the validity range. We can ignore it like the term aW_{ma} of dW/dt. Furthermore, the function $-dW/dt$ = f(W_{ma}) graph (on the bottom right of Figure 9.23) shows a very weak variation to this one with a maximum at the upper limit of the validity range. Therefore, by only retaining the term bW_{ma}, we have by integration between W_{ma} and $W_{ma} = 0$, the following expression for k_{ps} (equation 8.36):

$$k_{ps} = \frac{bW_{ma}}{\rho_w \bar{V} g} \approx -\frac{bW_{ma}}{\rho_w \bar{V} k_1} Exp(-\alpha_1 W_{ma}),$$

where b (negative) is in $kg_{water}/kg_{soil}/s$; k_1, like g and g_o, is in $kg_{water}/kg_{soil}/dm$.

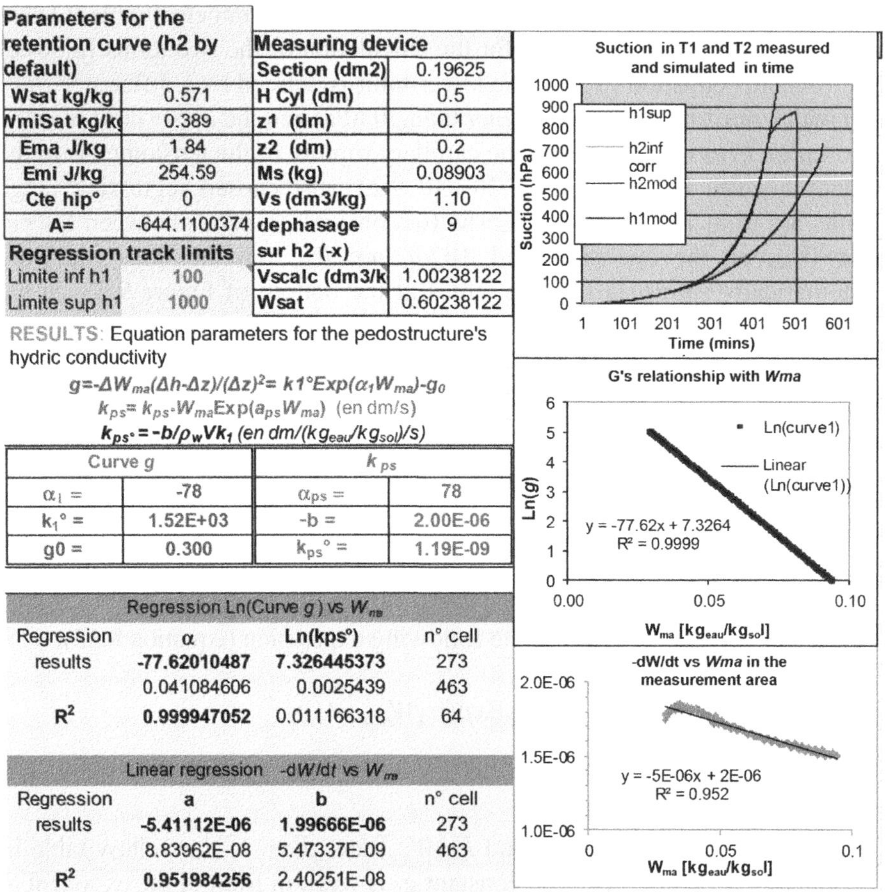

Parameters for the retention curve (h2 by default)	
Wsat kg/kg	0.571
WmiSat kg/k	0.389
Ema J/kg	1.84
Emi J/kg	254.59
Cte hip°	0
A=	-644.1100374
Regression track limits	
Limite inf h1	100
Limite sup h1	1000

Measuring device	
Section (dm2)	0.19625
H Cyl (dm)	0.5
z1 (dm)	0.1
z2 (dm)	0.2
Ms (kg)	0.08903
Vs (dm3/kg)	1.10
dephasage sur h2 (-x)	9
Vscalc (dm3/k	1.00238122
Wsat	0.60238122

RESULTS: Equation parameters for the pedostructure's hydric conductivity

$g = -\Delta W_{ma}(\Delta h - \Delta z)/(\Delta z)^2 = k1°Exp(\alpha_1 W_{ma}) - g_0$

$k_{ps} = k_{ps°} W_{ma} Exp(a_{ps} W_{ma})$ (en dm/s)

$k_{ps°} = -b/\rho_w V k_1$ (en dm/(kg$_{eau}$/kg$_{sol}$)/s)

Curve g		k_{ps}	
α_1 =	-78	α_{ps} =	78
$k_1°$ =	1.52E+03	-b =	2.00E-06
g0 =	0.300	$k_{ps}°$ =	1.19E-09

Regression Ln(Curve g) vs W_{ma}			
Regression results	α	Ln(kps°)	n° cell
	-77.62010487	**7.326445373**	273
	0.041084606	0.0025439	463
R^2	**0.999947052**	0.011166318	64

Linear regression -dW/dt vs W_{ma}			
Regression results	a	b	n° cell
	-5.41112E-06	**1.99666E-06**	273
	8.83962E-08	5.47337E-09	463
R^2	**0.951984256**	2.40251E-08	

Figure 9.23. *Graphs and tables resulting from the hydric conductivity calculation sheet. Halloysite soil from the B Ponterre soil parcel, Lézarde Riviera, Martinique (482). For a color version of the figure, see www.iste.co.uk/braudeau/hydro.zip*

In the "Results" table (Figure 9.23), we find

$\alpha_{ps} = -\alpha_1 = 78$ kg_{soil}/kg_{water}, $b = -2.0\ 10^{-6}$ $kg_{water}/kg_{soil}/s$

and

$W_{ma^\circ} = 0.14$ kg_{water}/kg_{soil}.

and

$k_{ps}^0 = -b/(\rho_w \bar{V} k_1) = 1.19\ 10^{-9}$ $dm/(kg_{water}/kg_{soil})/s$.

Conclusion

This book presented a unified and universal methodology for the characterization and systemic modeling of a natural organization, the soil, by developing the notions of General System and systemic description repository of organized objects. These notions, as well as the physics of the natural environment and their results (the thermodynamics and hydrodynamics of the soil medium) paves the way to the emergence of a new discipline in soil science, namely *hydrostructural pedology*. This new field of science occupies a central spot among the agro-environmental disciplines for which, generally, the study strongly depends on the hydrostructural state of the soil and its hydrical regime. These disciplines, therefore, can only be enhanced by the hydrostructural characterization of the soil and its thermodynamic modeling with water offered by hydrostructural pedology.

Water is the primordial element in the agro-environmental systems, not only as a mobile constituent, but also as the primary exchange material between them. The thermodynamic and hydrostructural physics of the soil-water interface developed here is the only one which allows for the physical and mechanistic coupling of biological processes with the pedoclimatic and hydrostructural dynamic of the soil, that the Kamel® model is able to simulate with accuracy. This model can be considered as a complete representation of the systemic paradigm of the hydrostructural pedology in which the key element is the pedostructure, the characterization and modeling tool: Kamel® and the end product: the SIRS-Soils (Geo Referenced Information System of Soils), basic layer of the NEO-GIS (Natural Environment Organization-Geographic Information System)

A new strategy for modeling agro-environmental systems should henceforth be implemented. Indeed, the concerned agro-environmental disciplines will necessarily adopt the description and modeling of the natural environment brought by hydrostructural pedology, in order to carry out the coupling of their biological, agronomical, geophysical models, with the organized soil medium. The Kamel® model, which models the hydrostructural functioning of the pedon, as representative of a soil mapping unit, is consequently at the heart of all these couplings.

Due to a lack of knowledge about soil as a structured and organized physical medium which is always in thermodynamic interaction with water, there has not been so far a functional and quantitative soil typology established, that describes it as natural physical medium supporting biological life. In pedology, the existing soil typology is mainly qualitative as it is uniquely built upon the morphological description of the soil organizations and the study of their mineralogical composition. Pedology as such lacks the hydrostructural description axis describing the hydric functioning of these organizations to complement the qualitative description with quantitative modeling of their activity. The present study fills in this gap by presenting a unified methodology for the measurement, characterization and calculation of the soil hydrostructural parameters and characteristic of the pedostructure. Thus, a complete typology of the pedostructures and the representative pedons, differentiating them by their three descriptive axes. This task should be undertaken for at least two essential reasons:

– to be able to generalize the results of all laboratories experimenting with the coupling of physical, biological or geo-chemical processes, with the soil medium, on a soil sample (undisturbed sample from the soil structure in place) which will have been characterized and identified by this new methodology;

– to update and bring back the existing soil maps and other pedological information in this physical and systemic approach of the natural organizations in order to augment them with geo-referenced soil information systems (SIRS-Soils). These SIRS Soils constitute the information basis and modeling stand/support for each expert system to be produced on a given agro-environmental subject.

Bibliography

[AHU 06] AHUJAT L.R., MA L., TIMLIN D., "Trans-disciplinary soil Physics research critical to synthesis and modeling of agriculture systems", *Soil Sci. Soc. Am. J.*, no. 70, pp. 311–326, 2006.

[ANG 00] ANGULO-JAMARILLO R., VANDERVAERE J.P., ROULIER S. *et al.*, "Field measurement of soil surface hydraulic properties by disc and ring infiltrometers: a review and recent developments", *Soil Tillage Research,* no. 55, pp. 1–29, 2000.

[ASS 14] ASSI A.T., ACCOLA J., HOVHANNISSIAN G. *et al.*, "Physics of the soil medium organization, part 2: pedostructure characterization through measurement and modeling of the soil moisture characteristic curves", *Front. Environ. Sci.*, vol. 2, no. 5, p. 17, 2014.

[BAL 06] BALSER T.C., MCMAHON K.D., BART D. *et al.*, "Bridging the gap between micro- and macro-scale perspectives on the role of microbial communities in global change ecology", *Plant Soil*, no. 289, pp. 59–70, 2006.

[BEA 72] BEAR J., *Dynamics of fluid in porous media*, American Elsevier, New York, 1972.

[BEL 08] BELHOUCHETTE H., BRAUDEAU E., HACHICHA M. *et al.*, "Integrating spatial soil organization data with a regional agricultural management simulation model: a case study in northern Tunisia", *Transactions of the ASABE*, no. 51, pp. 1099–1109, 2008.

[BEL 13] BELLIER G., BRAUDEAU E., Device for measurement coupled with water parameters of soil, WO 2013/004927 A1, World Intellectual Property Organization, WIPO, 2013.

[BER 68] BERTALANFFY L., *General System Theory – Foundations, Development, Applications*, Braziller, New York, 1968.

[BOR 02] BORDIN P., *SIG Concepts, Outils et Données*, Hermès-Lavoisier, Paris, 2002.

[BOI 04] BOIVIN P., GARNIER P., TESSIER D., "Relationship between clay content, clay type, and shrinkage properties of soil samples", *Soil Science Society of American Journal*, no. 68, pp. 1145–1153, 2004.

[BOI 06] BOIVIN P., SCHÄFFER B., TEMGOUA E. *et al.*, "Assessment of soil compaction using soil shrinkage modelling: experimental data and perspectives", *Soil and Tillage Research*, no. 88, pp. 65–79, 2006.

[BOU 99] BOUMA J., STOORVOGEL J., VAN ALPHEN B.J. *et al.*, "Precision agriculture and the changing paradigm of Agricultural research", *Soil Sci. Soc. Am. J.*, no. 63, pp. 1763–1768, 1999.

[BRA 87] BRAUDEAU E., "Mesure automatique de la rétraction d'échantillons de sol non remaniés", *Science du Sol*, no. 25, pp. 85–93, 1987.

[BRA 88a] BRAUDEAU E., "Équation généralisée des courbes de retrait d'échantillons de sol structurés", *Comptes Rendus de l'Académie des Sciences de Paris*, no. 307, pp. 1731–1734,1988.

[BRA 88b] BRAUDEAU E., "Essai de caractérisation quantitative de l'état structural d'un sol base sur l'étude de la courbe de retrait", *Comptes Rendus de l'Académie des Sciences de Paris*, no. 307, pp. 1933–1936, 1988.

[BRA 93] BRAUDEAU E., BRUAND A., "Détermination de la courbe de retrait de la phase argileuse à partir de la courbe de retrait établie sur échantillon de sol non remanié : application à une séquence de sols de Côte d'Ivoire", *Comptes-Rendus de l'Académie des Sciences de Paris*, no. 316, pp. 685–692, 1993.

[BRA 99] BRAUDEAU E., COSTANTINI J., BELLIER G. *et al.*, "New device and method for soil shrinkage curve measurement and characterization", *Soil Science Society of America Journal*, no. 63, pp. 525–535, 1999.

[BRA 01] BRAUDEAU E., ZIDI C., LOUKIL A. *et al.*, "Un système d'information pédologique, le SIRS-Soils du périmètre irrigué de Cébala-Borj-Touil. (Basse Vallée de la Majerda)", *Bulletin Sols de Tunisie*, numéro spécial 2001, Direction des sols, Tunis, Tunisie, 2001.

[BRA 04a] BRAUDEAU, E., MOHTAR R.H., "Water potential in non-rigid unsaturated soil-water medium", *Water Resources Research*, no. 40, 2004.

[BRA 04b] BRAUDEAU E., FRANGI J.P., MOTHAR R.H., "Characterizing non-rigid dual porosity structured soil medium using its shrinkage curve", *Soil Science Society of American Journal*, no. 68, pp. 353–370, 2004.

[BRA 05] BRAUDEAU E., SENE M., MOHTAR R.H., "Hydrostructural characteristics of two African tropical soils", *European Journal of Soil Science*, no. 56, pp. 375–388, 2005.

[BRA 06a] BRAUDEAU E., Kamel® Simile version. Agence pour la protection des programmes, Paris, 2006.

[BRA 06b] BRAUDEAU E., MOHTAR R.H., "Modeling the swelling curve for packed soil aggregates using the pédostructure concept", *Soil Science Society of America Journal*, no. 70, pp. 494–502, 2006.

[BRA 07] BRAUDEAU E., Le SIRS-Soils de la Martinique, Fianal report of the IRD/ PRAM 2005-2006 project, Horizon documentation-IRD, available at: http:// horizon. documentation.ird.fr/exl-doc/pleins_textes/divers12-11/010050248.pdf, 2007.

[BRA 09a] BRAUDEAU E., MOHTAR R.H., "Modeling the soil system: bridging the gap between pedology and soil-water physics", *Global Planet. Change J.*, no. 67, pp. 51–61, 2009.

[BRA 09b] BRAUDEAU E., MOHTAR R.H., EL GHEZAL N. *et al.*, "A multi-scale "soil water structure" model based on the pédostructure concept", *Hydrology and Earth System Sciences Discussions,* no. 6, pp. 1111–1163, 2009.

[BRA 13] BRAUDEAU,E., ASSI, A., ACCOLA, J. *et al.*, *TypoSoil User Manual – For Soils Hydrostructural Characterization*, Researchgate, 2013.

[BRA 14a] BRAUDEAU E., MOHTAR R.H., "A framework for soil-water modeling using the pédostructure and structural representative elementary volume (SREV) concepts", *Front. Environ. Sci.*, no. 2, p. 24, 2014.

[BRA 14b] BRAUDEAU E., MOHTAR R.H., "Integrative environmental modeling", *Encyclopedia of agricultural, food, and biological engineering,* 2nd edition, 2014.

[BRA 14c] BRAUDEAU E., ASSI A.T., BOUKCIM H. *et al.*, "Physics of the soil medium organization, part 1 : thermodynamic formulation of the pédostructure water retention and shrinkage curves", *Front. Environ. Sci.*, vol. 2, article 4, p. 17, 2014.

[BRA 14d] BRAUDEAU E., HOVHANNISSIAN G., ASSI A.T. *et al.*, "Soil water thermodynamic to unify water retention curve by pressure plates and tensiometer", *Front. Earth Sci*, vol. 2, article 30, p. 13, 2014.

[BRE 64] BREWER R., *Fabric and mineral analysis of soils*, John Wiley and Sons, New York, 1964.

[BRE 08] BELHOUCHETTE H., BRAUDEAU E. *et al.*, "Integrating spatial soil organization data with a regional agricultural management simulation model : a case study in northern Tunisia", *Transactions of the ASABE*, no. 51, pp. 1099–1109, 2008.

[COL 96] COLLEUILLE H., BRAUDEAU E., "A soil fractionation related to soilstructure behaviour", *Australian Journal of Soil Research*, no. 34, pp. 653–669, 1996.

[CPC 67] CPCS (COMMISSION DE PEDOLOGIE ET DE CARTOGRAPHIE DES SOLS), Classification des sols. Grignon, France : Laboratoire de géologie et de pédologie, Ecole nationale supérieure d'agronomie, p. 87, available at: http://www.afes.fr/afes/docs/CPCS_1967.pdf, 1967.

[DEE 15] DEEB M., Influence des plantes, des vers de terre et de la matière organique sur la structure de technosols construits, thesis, Université Paris Est, 2015.

[DER 01] DEROUICHE C., LOUKIL A., DECLUSEAU D. *et al.*, "Le SIRS-SOILS du périmètre irrigué de Cébala", *Bulletin Sols de Tunisie*, Tunis, Tunisia, pp. 51–66, 2001.

[DON 10] DONATELLI M., RUSSELL G., RIZZOLI A.E. *et al.*, "A Component-Based Framework for Simulating Agricultural Production and Externalities", in BROUWER F., VAN ITTERSUM M.K. (eds), *Environmental and agricultural modelling : integrated approaches for policy impact assessment*, Springer, Dordrecht, Holland, no. 581, pp. 63–108, 2010.

[DUC 77] DUCHAUFOUR P., SOUCHIER B., *Pédogenèse Et Classification*, Masson, Paris, 1977.

[KLI 72] KLIR G., *Trends in General Systems Theory*, John Wiley, New York, 1972.

[LEM 94] LE MOIGNE J.-L., *La théorie du système général, Théorie de la modélisation*, 4th edition, PUF, Paris, 1994.

[LEO 97] LEONG E.C., RAHARDJO H., "Review of soil water characteristic curve equations", *Journal of Geotechnical and Geoenvironmental Engineering*, vol. 123, no. 12, pp. 1106–1117, 1997.

[LIN 03] LIN H., "Hydropedology: Bridging Disciplines, Scales, and Data", *Vadoze Zone J.*, no. 2, pp. 1–11, 2003.

[LIN 06] LIN H., BOUMA H., PACHEPSKY Y. *et al.*, "Hydropedology: Synergistic integration of pedology and hydrology", *Water Resour. Res.*, no. 42, 2006.

[LIN 12] LIN H., *Hydropedology: Synergistic Integration of Soil Science and Hydrology*, 1st edition, Elsevier, Amsterdam, 2012.

[MES 64] MESAROVIĆ M.D., "Foundations for a general systems theory", in MESAROVIĆ M.D. (ed), *Views on General Systems Theory*, Wiley, New York, pp. 1–24, 1964.

[PAC 03] PACHEPSKY Y.A., RAWLS W.J. "Soil structure and pedotransfer functions", *European Journal of Soil Science*, no. 54, pp. 443–451, 2003.

[POU 13] POUVREAU D., Une histoire de la systémologie générale de Ludwig von Bertalanffy. Généalogie, genèse, actualisation et postérité d'un projet herméneutique", PhD Thesis, Ecole des hautes études en sciences sociales (EHESS), Paris, available at: http://tel. archives- ouvertes.fr/tel-00804157, 2013.

[SAL 12] SALAHAT M., MOHTARR. H., BRAUDEAU E. *et al.*, "Toward delineating hydrofunctional soil mapping units using the pedostructure concept: A case study", *Computers and Electronics in Agriculture*, no. 86, pp. 15–25, 2012.

[SAX 06] SAXTON K.E., RAWLS W.J., "Soil water characteristic estimates by texture and organic matter for hydrologic solutions", *Soil Sci. Soc. Am. J.*, no. 70, pp. 1569–1578, 2006.

[SCH 10] SCHINDLER U., DURNER W., VON UNOLD G. *et al.*, "Evaporation method for measuring unsaturated hydraulic properties of soils: extending the measurement range", *Soil Science Society of America*, vol. 74, no. 4, pp. 1071–1083, 2010.

[SIN 12] SINGH J., MOHTAR R., BRAUDEAU E. *et al.*, "Field evaluation of the pedostructure-based model (Kamel®)", *Comput. Electron. Agric*, 2012.

[SPO 81] SPOSITO G., *The Thermodynamics of Soil Solutions*, Oxford University Press, New York, 1981.

[US 93] US DEPARTMENT OF AGRICULTURE, *Soil Survey Division Staff, Soil Survey Manual,* Soil Conservation Service, Handbook 18, available at: http://soils.usda.gov/technical/manual, 1993.

Index

A, B, C

D, F, G

H, I, M

N, O, P

R, S, T

Other titles from

iSTE

in

Earth Systems – Environmental Engineering

2016

QUEVAUVILLER Philippe
Marine Chemical Monitoring

2015

HERBIN Hervé, DUBUISSON Philippe
Infrared Observation of Earth's Atmosphere

MERCIER-LAURENT Eunika
The Innovation Biosphere

MONACO André, PROUZET Patrick
Seas and Oceans Set
Volume 7 – Marine Ecosystems: Diversity and Functions
Volume 8 – Governance of Seas and Oceans

2014

DE LARMINAT Philippe
Climate Change: Identification and Projections

MONACO André, PROUZET Patrick
Seas and Oceans Set
Volume 1 – Ocean in the Earth System
Volume 2 – The Land-Sea Interactions
Volume 3 – Ecosystem Sustainability and Global Change
Volume 4 – Vulnerability of Coastal Ecosystems and Adaptation
Volume 5 – Development of Marine Resources
Volume 6 – Value and Economy of Marine Resources

MUTTIN Frédéric
Marine Coastal and Water Pollutions: Oil Spill Studies

2013

LÉCUYER Christophe
Water on Earth: Physicochemical and Biological Properties

LEGCHENKO Anatoly
Magnetic Resonance Imaging for Groundwater

2012

LE MENN Marc
Instrumentation and Metrology in Oceanography

PAVÉ Alain
Modeling of Living Systems: From Cell to Ecosystem

2010

TANGUY Jean-Michel
Environmental Hydraulics Series
Volume 1 – Physical Processes and Measurement Devices
Volume 2 – Mathematical Models
Volume 3 – Numerical Methods
Volume 4 – Practical Applications in Engineering
Volume 5 – Modeling Software

Printed and bound by CPI Group (UK) Ltd, Croydon, CR0 4YY
13/09/2026
14961054-0003